D1085117

DEC 2008

THE UNFINISHED GAME

The Unfinished Game

Pascal, Fermat, and the
Seventeenth-Century Letter that
Made the World Modern

KEITH DEVLIN

A Member of the Perseus Books Group
New York

Copyright © 2008 by Keith Devlin

Published by Basic Books,
A Member of the Perseus Books Group

All rights reserved. Printed in the United States of America. No part of this book may be reproduced in any manner whatsoever without written permission except in the case of brief quotations embodied in critical articles and reviews. For information, address Basic Books, 387 Park Avenue South, New York, NY 10016-8810.

Books published by Basic Books are available at special discounts for bulk purchases in the United States by corporations, institutions, and other organizations. For more information, please contact the Special Markets Department at the Perseus Books Group, 2300 Chestnut Street, Suite 200, Philadelphia, PA 19103, or call (800) 810-4145, ext. 5000, or e-mail special.markets@perseusbooks.com.

Designed by Brent Wilcox

Library of Congress Cataloging-in-Publication Data
Devlin, Keith J.
    The unfinished game : Pascal, Fermat, and the seventeenth-century letter that made the world modern / Keith Devlin.
        p.   cm.
    Includes bibliographical references and index.
    ISBN 978-0-465-00910-7 (alk. paper)
    1. Probabilities.   2. Pascal, Blaise, 1623–1662—Correspondence.   3. Fermat, Pierre de, 1601–1665—Correspondence.   4. Mathematicians—France—Correspondence.
I. Title.
QA273D455 2008
519.2—dc22

                                                  2008012222

10 9 8 7 6 5 4 3 2 1

519.2 D497u 2008
Devlin, Keith J.
The unfinished game : Pascal
Fermat, and the

CONTENTS

NOTE TO THE READER

The Pascal-Fermat correspondence is displayed and set in italics for easy identification, to facilitate checking back to follow the entire thread. (The main letter is revealed chapter by chapter.) Other letters in the Pascal-Fermat correspondence are also set in italics. Displayed quotations from others are set in small Roman font.

PREFACE

When Basic Books editor Bill Frucht approached me with the idea of a book about a single mathematics document that changed the course of history, my first thought was the letter Pascal wrote to Fermat in the seventeenth century that established modern probability theory. Everyone who has taken a course in probability theory will have heard the story of this letter. But few people, myself included, have ever read the original or looked into the story that surrounds it. Eager to work with Bill once again (he edited my earlier books *The Math Gene* and *The Millennium Problems*), I was moved to look more closely at the famous correspondence. What I discovered astonished me. The story is far more fascinating than I knew, and the letter itself provides a revealing insight into how two of the world's greatest mathematical minds grappled with a problem, stumbling, making "elementary" errors, eventually being rewarded with flashes of brilliance. What also became clear from reading what the experts wrote before and after this letter was just how great was the revolution it brought about in the way humans think about the future. Much of the way we live our lives today was made possible by the mathematical ideas expressed in those few pages.

In addition to Bill, who has once again been a delight to work with, this project benefited from the editorial hand of Basic's Courtney Miller and from the advice and guidance of my agent, Ted Weinstein. Professor John Stillwell, who read the first draft of the manuscript, put my mind at rest that—as a mathematician, not a historian of mathematics—I had not made any serious gaffes in reporting the historical development.

Keith Devlin
Palo Alto, California

CHAPTER 1

Monday, August 24, 1654

Monsieur,

1. I was not able to tell you my entire thoughts regarding the problem of the points by the last post, and at the same time, I have a certain reluctance at doing it for fear lest this admirable harmony which obtains between us and which is so dear to me should begin to flag, for I am afraid that we may have different opinions on this subject. I wish to lay my whole reasoning before you, and to have you do me the favor to set me straight if I am in error or to indorse me if I am correct. I ask you this in all faith and sincerity for I am not certain even that you will be on my side. °

°The original was written in French. The translation presented here is by Professor Vera Sanford, Western Reserve University, Cleveland, and appears in D. E. Smith, ed., *A Source Book in Mathematics* (Mineola, NY: Dover, 1959). Smith's volume is the source of all excerpts from the letters between Pascal and Fermat in the pages that follow. Editorial inserts are those of Sanford. Pascal numbered the sections of his letter for later reference. The letter is revealed chapter by chapter as our story unfolds. For quick reference, the letter is reproduced in its entirety at the end of the book.

On Monday, August 24, 1654, the famous French mathematician Blaise Pascal sat at his desk and composed a letter to his equally famous countryman Pierre de Fermat. When completed, the letter would come to less than three thousand words, but it would change human life forever. It set out, for the first time, a method whereby humans can predict the future.

Not in the sense of saying what *will* happen; no one can do that, not even if you are Blaise Pascal, a child prodigy who at sixteen wrote his first original paper and while still a teenager had invented, manufactured, and sold a mechanical desk calculator (the Pascaline) that largely anticipated the commercial desk calculators that became commonplace some three hundred years later. Rather, Pascal's letter showed how to predict the future by calculating, often with extraordinary precision, the numerical *likelihood* of a particular event's occurring.

Within a few years of Pascal's sending his letter, people no longer saw the future as completely unpredictable and beyond their control. They could compute the likelihoods of various things' happening and plan their activities—and their lives—accordingly. In short, Pascal showed us how to manage risk. His letter created our modern view of the future.

RISK

Today, we are so used to the idea that the future comes with probabilities—the chance that it will rain tomorrow, that house prices will rise over the coming months, that an individual will die within the next five years, or that terrorists will attack Los Angeles International Airport—that it is hard to imagine life being any other way. We insure our lives, homes, automobiles, and vacations and find it unremarkable that a company can make money by selling us that insurance. The engineers who design and build the aircraft we fly in know that their products are flawed, but because they can calculate with enormous precision the likelihood that a flaw will cause a major crash and determine that the risk is acceptably low, they are able to make the crucial decision to manufacture the plane and put it into commission. (NASA carried out essentially the same calculation in designing the space shuttle and concluded that a major disaster was likely in roughly one out of every two hundred flights, a figure it judged to be an acceptable risk. The calculation has proved tragically accurate. The risk of a Boeing 777 crashing is far, far less, but the companies that insure aircraft know it with considerable precision.)

Once humanity had a way to calculate the probabilities of future events, we were able to plan our lives with far greater precision than formerly. Not just scientists and engineers but also ordinary people were suddenly able to do things and go places that hitherto had been the province of the reckless, the rich, or the ignorant. Business, politics, defense, warfare, science, engineering, medicine, sport, recreation, finance, housing, transportation, and many other aspects of everyday life are today filled with probability computations.

Even those who are not schooled in the mathematics of calculating odds know that the future is not a matter of blind fate. We can often judge what is likely to happen and plan accordingly. Yet before Pascal wrote his letter to Fermat, many learned people (including some leading mathematicians) believed that predicting the likelihood of future events was simply not possible.

The ability to calculate probabilities transformed the practice of statistics, changing it from the mere collection and tabulation of data to the use of data to draw inferences and make informed decisions. Without the ability to quantify risk, there would be no liquid capital markets, and global companies like Google, Yahoo!, Microsoft, DuPont, Alcoa, Merck, Boeing, and McDonald's might never have come into being. The pundits and pollsters who today tell us who is likely to win the next election make direct use of mathematical techniques developed by Pascal and Fermat. In modern medicine, future-predictive statistical methods are used all the time to compare the benefits of various drugs and treatments with their risks. As a result, the majority of people in the developed world now lead far longer and healthier lives than at any previous time in history.

Within a hundred years of Pascal's letter, life-expectancy tables formed the basis for the sale of life annuities in England, and London was the center of a flourishing marine insurance business, without which sea transportation would have remained a domain only for those who could afford to assume the enormous risks it entailed. Still another legacy of that correspondence is the casino industry. It truly is an industry, and a highly profitable one, because the casinos do not gamble—they leave that to their customers. Using the ability that probability theory gives them to predict the outcomes of games of chance, they can calculate in advance exactly how much money they will make each week. Political pollsters sometimes get the answer wrong; for them, predicting the future is not 100 percent certain. But the casinos never get it wrong.

The list could go on. Managing risk is now fundamental to almost every aspect of our lives. It's part of the way we see life, and we take it for granted. Yet when you read the letter from Pascal to Fermat and see the enormous difficulty these two great mathematicians had in grasping the very *idea* of predicting the likelihood of future events, let alone *how* to do so, you realize that what we nowadays take for granted was a huge advance in human thinking that came only through significant intellectual effort. Others had tried to achieve the same breakthrough, and failed.

A LONG STRUGGLE

Risk has always held a fascination, and games of chance are as old as civilization. Excavations from the pyramids have uncovered *astralagi* (dice crafted from the ankle bones of animals) used by the pharaohs. Some of these dice were loaded, showing that the desire to improve one's chances by cunning or subterfuge was there from the beginning, even though the Egyptians did not have the mathematics to guide their play. Ancient Greek vases have drawings showing young men tossing bones into a circle. Pontius Pilate is said to have taken bids for Christ's robe during the crucifixion. The sandwich was invented and named for the Earl of Sandwich, who wanted to take his meals without having to leave the gaming table. George Washington famously gambled in his tent during the American Revolution.

The fascination with chance also permeates our myths and legends. Greek mythology has the gods on Mount Olympus rolling dice to determine men's fates. And three brothers rolled dice for the universe, with Zeus winning the heavens, Poseidon the seas, and Hades, the loser, condemned to be master of the underworld.

Yet before the Middle Ages, no one suspected it was possible to quantify the likelihood of some future event such as the outcome of a roll of dice. Events that were not in some way predetermined were thought to be beyond rational analysis. The future is a matter determined by God, people felt; what will happen will happen, and there is nothing anyone can do about it other than pray and show allegiance to whatever deity you believe controls your destiny.

Aristotle, to whom we owe so much of our mathematics and philosophy, taught that there were just three kinds of events: (1) certain events, which occur necessarily, such as the next sunrise; (2) probable events, which happen most of the time, such as its being warm and sunny at midday in

July (in California, where I live); and (3) unpredictable events, which happen by pure chance and are thus unknowable. Games of chance clearly fall into Aristotle's third category, and there was no thought that someone could play so as to improve his chance of winning (other than by cheating). Neither Aristotle nor any of his compatriots seems to have noticed that there are certain numerical patterns to the outcomes when the same game is played repeatedly—that some outcomes occur more frequently than others (such as getting a 7, compared with a 6, when two dice are rolled). This may seem surprising in view of the Greeks' great prowess in mathematics and science, but since they did not use precisely and uniformly manufactured randomizing devices in their games, perhaps there were no regular patterns to be discerned. For instance, games involving throwing dice were very popular, but the Greeks did not roll our familiar, regular cubic dice. Instead, like the pharaohs, they played with *astralagi*, which had two rounded sides and only four playable surfaces, no two of which were identical. Although the Greeks did seem to believe that certain throws were more likely than others, these beliefs—superstitions—were not based on observation, and some were at variance with the actual likelihoods we would calculate today. Likewise, although the Roman emperors used lotteries as a way to raise funds, there was no attempt to analyze them mathematically.

The first known attempt to discern numerical patterns in games of chance seems to have come around 960, when Bishop Wibold of Cambrai correctly enumerated the 56 outcomes that can arise when three dice are thrown simultaneously: 1, 1, 1; 1, 1, 2; 2, 3, 5; and so on. A thirteenth-century Latin poem, *De Vetula*, listed the 216 (= 6 × 6 × 6) outcomes that may result when three dice are thrown in succession.

During the fourteenth century, dice games were supplemented by card games, which were first seen in Europe in Italy, though objections from the Church hindered their growth. Still, no attempt was made to tabulate and analyze the frequencies of various hands until the start of the fifteenth century, when Italian mathematicians began to look for ways to improve the chances of winning at the gaming tables, sometimes for themselves, other times at the behest of wealthy noblemen.

In 1494, in his book *Summa de arithmetica, geometrica, proportioni et proportionalita (Everything About Arithmetic, Geometry, and Proportions)*, Luca Pacioli first put into print the problem that Pascal and Fermat would solve two centuries later. Known as the *problem of the unfinished game*, it asks how the stakes should be divided when a game of several rounds must be abandoned unfinished. The puzzle is also known as the *problem of the points*, since rather than counting rounds, we can assign each player points for winning each round. The exact origin of the puzzle is unknown, but it seems to predate Pacioli's reference to it. This is the problem Pascal refers to in the opening of his letter.

THE PROBLEM OF THE POINTS

Suppose two players—call them Blaise and Pierre—place equal bets on who will win the best of five tosses of a fair coin. They start the game, but then have to stop before either player has won. How do they divide the pot?

If each has won one toss when the game is abandoned after two throws, then clearly, they split the pot evenly, and if they abandon the game after four tosses when each has won twice, they do likewise. But what if they stop after three tosses, with one player ahead 2 to 1?

This is a particularly simple version of the problem, not exactly the form Pascal and Fermat considered. But the logic required for its solution is the same as that for the problem as originally formulated, which was in terms of throwing dice and was in a general form that allowed for "best of N throws" for any number N. (When Pascal and Fermat were working on the problem, they did, however, simplify the situation to the case where the dice were equivalent to coins, having, say, 0 on three faces and 1 on the other three, and they did look at particular, small values of N.)

The key to solving the problem of the points is to find a way to look into the future—the future as it would have (or might have) unfolded had the two players continued the game. On the face of it, this is impossible. How can anyone possibly know how things would have turned out? It took two of the world's greatest mathematical geniuses several weeks of intense intellectual effort to see how to get past this seeming impasse.

The sheer difficulty of the task is clear from Pascal's uncertainty about his own reasoning. His opening paragraph lays out this uncertainty:

> I wish to lay my whole reasoning before you, and to have you do me the favor to set me straight if I am in error or to indorse me if I am correct. I ask you this in all faith and sincerity for I am not certain even that you will be on my side.

Pascal continued:

> When there are but two players, your theory which proceeds by combinations* is very just. But when there are three, I believe I have a proof that it is unjust that you should proceed in any other manner than the one I have. But the method which I have disclosed to you and which I have used universally is common to all imaginable conditions of all distributions of points, in the place of that of

---

*By combinations he means listing all the possible ways, or combinations of dice outcomes, that could have arisen if the game had continued.

combinations (which I do not use except in particular cases when it is shorter than the general method), a method which is good only in isolated cases and not good for others.

I am sure that I can make it understood, but it requires a few words from me and a little patience from you.

Even the experts can find it difficult to master a new mathematical idea.

CHAPTER 2

A Problem Worthy of Great Minds

2. This is the method of procedure when there are two play-ers. If two players, playing in several throws, find them-selves in such a state that the first lacks two points and the second three of gaining the stake, you say it is necessary to see in how many points the game will be absolutely decided.

It is convenient to suppose that this will be in four points, from which you conclude that it is necessary to see how many ways the four points may be distributed between the two players and to see how many combina-tions there are to make the first win and how many to make the second win, and to divide the stake according to that proportion. I could scarcely understand this reason-ing if I had not known it myself before; but you also have written it in your discussion.

Pascal continues his letter, following his practice of num-bering each section for subsequent reference. Since the two mathematicians have already exchanged several letters on the subject, in his second section, he can dive straight into the details.*

PROBABILITY

Today, we would use the word *probability* to refer to the focus of Pascal and Fermat's discussion, but that term was not introduced until nearly a century after the mathematicians' deaths. Instead, they spoke of "hazards," or number of chances. Much of their difficulty was that they did not yet have the notion of mathematical probability—because they were in the process of inventing it.

From our perspective, it is hard to understand just why they found it so difficult. But that reflects the massive change in human thinking that their work led to. Today, it is part of our very worldview that we see things in terms of probabilities.

Yet, for all its familiarity, probability remains a tricky notion to deal with. For one thing, there are actually several notions of probability. Two in particular are common: frequentist probability and subjective probability.**

---

*The entire correspondence took place in 1654. It is not known how many letters were involved. Seven are extant today: three from Pascal and four from Fermat, spanning the period July to October. They can all be found in the volume *Oeuvres de Fermat*, vol. 2 (1894). English translations are provided in D. E. Smith, *A Source Book in Mathematics* (New York: Dover, 1959).

**The distinction between different forms of probability was first made in 1713 by the Swiss mathematician Jakob Bernoulli, who is largely responsible for taking Pascal and Fermat's ideas and expressing them in what became the terminology of modern probability theory.

Frequentist probability is the one most people are familiar with today. It's the notion that arises in the classic games of chance such as cards, dice, or roulette, where you can use a theoretical mathematical argument to calculate the probability of a certain event, such as rolling boxcars (double 6's) when you throw two dice. (The probability in this case is 1/36.) It also occurs when you have some large population of people, animals, or objects, or some action that is repeated many times, where you can count the relative frequency of a particular outcome—say, the probability that a man chosen at random from a large population will die before he reaches sixty. The frequentist probability of an event E occurring in some population (or repeated action) A is the number of different ways E can occur (or the number of times E does occur) divided by the total number of different outcomes. For example, if you roll a die, it can land six different ways. There are three ways it can land with an even number face up (2, 4, or 6), so the probability of rolling an even number with an honest die is 3/6, or 1/2.

Subjective probability, also called *epistemic probability,* refers to a numerical estimate of the veracity of our knowledge of some event, where that knowledge is not based (entirely) on statistical data—for example, when a woman tells you she is 95 percent certain she knows the way to the post office.

This seemingly clear distinction becomes blurred when a subjective probability is based on data, but not in an entirely deterministic way, or when techniques such as Bayes' formula (which we'll get to later) are used to refine subjective probabilities in the light of concrete data. Although my primary focus in this book is on the development of frequentist probability, much of its impact on our world has come when its precise mathematical calculus has been used to base or to buttress subjective decision making.

But that was all in the future when Pascal and Fermat, with no idea of the consequences of their undertaking, embarked on their quest to solve the problem of the points.

Struggling Toward the Solution

Today, anyone who has had just a few hours of instruction in probability theory can solve the problem of the points with ease, either the simple version I gave in the last chapter or the one Pacioli considered or even, for someone comfortable with elementary algebra, the general case of "N rounds."*
But the original solution came only after much effort, during which time different mathematicians proposed quite different answers, of which only one could possibly be correct.

Pacioli, the man who first wrote about the problem, considered a version in which the game is played until one player has won six rounds, but play is abandoned when the score is 5 to 2. He suggested that the solution is to divide the pot according to the current state of play, namely, 5 to 2, but this reasoning is incorrect. The flaw in Pacioli's reasoning

---

*I shall present a complete solution to the simple case later in this chapter.

was demonstrated in 1539 by the next person to try to solve the problem, his countryman Girolamo Cardano.

Cardano noted, correctly, that the apportionment of the pot depended not on how many rounds each player had already won (as Pacioli thought) but on how many each player must still win in order to win the contest. Although this insight helped pave the way to the eventual solution, Cardano, too, failed to find the right answer. He did, however, make several key observations that established the beginning of what, following Pascal and Fermat's work, became probability theory.

A mathematician and a physician, Cardano published books on both disciplines during a highly colorful life. Because of his illegitimate birth and sharp tongue, he was initially denied admission to the College of Physicians in Milan, only to be admitted later in life because of his tremendous achievements. He worked as a country doctor, a lecturer in mathematics in Milan (in which city he grew to be the most prominent and most sought-after physician), and then as a professor of medicine in Bologna—a position from which he was subsequently dismissed after being arrested and accused of heresy. In addition to books on mathematics and medicine, Cardano also wrote on astronomy, physics, chess, death, the immortality of the soul, wisdom, and games of chance, the last reflecting another prominent side of his multifaceted life, his compulsive gambling.

Cardano's lasting contribution to the creation of probability theory came in a manuscript he wrote in 1564, titled

Liber de ludo aleae (The Book on Games of Chance) and published in 1663, almost a century after his death. He wrote it not as a mathematics treatise but as a practical guide for gamblers. Buried among many pages of advice for the gaming table is some important mathematics, in particular the rule for when odds may be added (namely, when a game splits into separate cases) and the derivation of the hugely important multiplication rule for combining odds when a game is repeated several times. (The first rule tells you that if you throw a die, the odds of getting a 1 or an even number are $1/6 + 1/2 = 2/3$; the second rule tells you that when you throw a die twice, the odds of getting a 1 followed by an even number are $1/6 \times 1/2 = 1/12$.)

The extent to which our modern approach to probability calculations is a recent innovation is made clear by the fact that not only would it be some time before Cardano's ideas were generally accepted, but another leading Italian mathematician, Niccolò Tartaglia, in joining Cardano in deriding Pacioli's solution in 1556, added that he believed the problem could not be solved at all! In 1603, another Italian mathematician, Lorenzo Forestani, reached essentially the same conclusion. In his book *Practica d'arithmetica e geometria (The Practice of Arithmetic and Geometry)*, he suggested that the portion of the stake should be divided based on the number each had won in relation to the number of games played, with the remainder divided equally between them, because the remainder of the game favors neither player. This is a hopeless analysis by today's standards, but, as with Tartaglia, it reflected the widely held belief that the future was a matter of pure chance, with every possibility equally likely.

The last major figure to enter the picture prior to Pascal and Fermat was no less than Galileo Galilei, the father of modern science, who wrote a paper sometime between 1613 and 1623 titled *Sopra le scoperte dei dadi (On a Discovery Concerning Dice)*, in which he analyzed (completely and correctly) all the ways three dice will give totals of 9 and 10 points. (Discounting order, each total can be obtained six different ways.) His motivation for doing this, it seems, was to justify the belief among gamblers that a total of 10 is a slightly better bet than 9. Galileo demonstrated that this is indeed the case, by calculating that a total of 10 can be obtained by way of twenty-seven different dice-throw outcomes, whereas 9 can be obtained by only twenty-five. This problem had also been considered and solved by Cardano. The significance of Galileo's contribution was that, whereas Cardano had reasoned theoretically, Galileo approached the problem scientifically, beginning with an empirical observation—that 10 occurs more often than 9 (the difference is sufficiently small that it takes a lot of plays and a very keen eye to spot this pattern)—and then seeking a mathematical explanation. This very much set the stage for what came next.

PASCAL AND FERMAT

Blaise Pascal (1623–1662) was a child prodigy. He was born on June 19 in Clermont, France, today's Clermont-Ferrand.

His mother died when he was three, and a few years later, Pascal's father, Étienne, a wealthy tax official and a keen amateur mathematician, moved the family from Clermont to Paris, where he personally oversaw his son's education at home.

Étienne maintained some odd views. He decided that his son should not study mathematics before the age of fifteen and, accordingly, removed all mathematics texts from their house. This prohibition only raised young Blaise's curiosity about the banned subject, and he started to work on geometry in secret at the age of twelve. He discovered on his own that the sum of the angles of a triangle is two right angles; when Étienne found out, he was so impressed that he removed the ban and allowed his son to read mathematics texts, starting with Euclid's classic work *Elements.* He also started to take the obviously gifted Blaise to meetings of Mersenne's Academy, one of several semiformal groups of mathematicians and scientists in Paris that eventually gave birth to the Académie Royale des Sciences in 1666. As I noted in Chapter 1, when Pascal was sixteen, he wrote his first paper, on conic sections, and presented it to Mersenne's Academy.

To assist his father's tax-collecting work, the teenage Pascal also invented the calculating machine I mentioned earlier, and oversaw its manufacture and sale. The device, the Pascaline, looked much like the mechanical calculators that were sold throughout the world in the 1940s and 1950s. Pascal worked on developing his calculator for three years, between 1642 and 1645. Perhaps because his interests lay elsewhere, he was not a success as a calculator entrepreneur: the device sold only in small numbers and eventually went out of production.

The adult Pascal devoted his life to the study of mathematics, natural science, and religion, all privately supported by his family's fortune (he never took a university position). One of his better-known mathematical studies was what we call Pascal's triangle. To generate the triangle, you start with a 1, and then immediately below it, you put two 1's, one to either side. Then for each successive row, you put a new 1 at either end and complete the row between them by adding together each adjacent pair of entries in the row above and putting their sum halfway between them. Here are the first few rows:

```
                    1
                1       1
            1       2       1
        1       3       3       1
    1       4       6       4       1
1       5       10      10      5       1
```

From today's perspective, Pascal's triangle does not appear mathematically deep; nor do the many interesting properties Pascal discovered that relate the various entries. But it turned out to be particularly important in elementary algebra and in probability theory, since the entries in each row are the so-called binomial coefficients that occur in the expansions of the expression $(a + b)^n$. For example

$$(a + b)^2 = a^2 + 2ab + b^2 \qquad\qquad [1\text{-}2\text{-}1]$$
$$(a + b)^3 = a^3 + 3a^2b + 3ab^2 + b^3 \qquad\qquad [1\text{-}3\text{-}3\text{-}1]$$

In his twenties, Pascal became ill and never fully regained his health (he died at age thirty-nine). In his later years, he lost interest in mathematics and focused his attention on writing religious treatises.

Pierre de Fermat (1601–1665) was born to a family of wealthy merchants. He studied law at the universities of Toulouse and Orléans, and mathematics at Bordeaux. His law studies prepared him for his subsequent career as a lawyer and jurist; his mathematical training prepared him for his lifelong interest in the subject. Although Fermat is often described today as an amateur mathematician, this is true only in the sense that he did not get paid for his research. He devoted immense amounts of time to it and was famed in his day as one of the greatest mathematicians in Europe. With no need to earn a living doing mathematics, he published almost nothing. Instead, he carried out his work and dispersed his ideas and results through ongoing correspondences with the leading mathematicians of the time.

Fermat worked in several areas of mathematics, including geometry, where he developed algebraic coordinate geometry independently of René Descartes, after whom it is often named. He also made important contributions to the early development of calculus. But he is known best for his research into number theory, the branch of mathematics that looks at properties of the positive whole numbers. With a distinguished history going back to the ancient Greeks, number theory was in Fermat's time regarded as one of the pinnacles of mathematics. (It remains so today.) Fermat made many profound discoveries in the field, though his most widespread fame came from an observation that (to the best of our knowledge) he never actually proved. He claimed that the equation

$$x^n + y^n = z^n$$

has no solution in which x, y, and z are all positive whole numbers and n is a whole number greater than 2. (The solutions when n is 2 are, of course, familiar to anyone who's taken high school geometry.) Fermat made a brief note in the margin of a book—Diophantus's *Arithmetica*—saying he had found a "truly marvelous proof" of this conjecture but that it was too big to fit in the margin. Discovered after his death, the conjecture eventually became known as Fermat's last theorem, the name reflecting the fact that of all the results he claimed during his mathematical career, this was the only one that no one was able to prove or disprove, even hundreds of years after his death. It was finally proved in 1994 by the English mathematician Andrew Wiles, by means of a long, complicated argument using techniques not available in Fermat's time. Most mathematicians believe Fermat was mistaken, and what he thought was a proof turned out to be flawed—which may be why he never revealed it.

THE STRATEGY

In 1654, the gambler Antoine Gombaud, whose noble title was the Chevalier de Méré,* approached his friend Pascal with some questions about games of chance, including the problem of the unfinished game. After some thought, Pascal found a possible solution but was not completely sure his reasoning was correct. Accordingly, he sent his ideas to Fermat to see if his countryman agreed with the argument. The brief exchange of letters that ensued—and one letter in particular—represented one of the most profound advancements in the history of mathematical thought.

Before we take a look at their exchange and the methods it contains, let's look at a present-day solution of the simple version of the problem. In this version, the players, Blaise and Pierre, place equal bets on who will win the best of five tosses of a fair coin. We'll suppose that on each round, Blaise chooses heads, Pierre tails. Now suppose they have to abandon the game after three tosses, with Blaise ahead 2 to 1. How do they divide the pot?

The idea is to look at all possible ways the game might have turned out had they played all five rounds. Since Blaise

---

*Chevalier was the lowest rank of nobility in France at the time, equivalent to a knight in England. The word is related to the English chivalry.

is ahead 2 to 1 after round three, the first three rounds must have yielded two heads and one tail.

The remaining two throws can yield

H H   H T   T H   T T

Each of these four is equally likely. In the first (H H), the final outcome is four heads and one tail, so Blaise wins; in the second and the third (H T and T H), the final outcome is three heads and two tails, so again Blaise wins; in the fourth (T T), the final outcome is two heads and three tails, so Pierre wins. This means that in three of the four possible ways the game could have ended, Blaise wins, and in only one possible play does Pierre win. Blaise has a 3-to-1 advantage over Pierre when they abandon the game; therefore, the pot should be divided 3/4 for Blaise and 1/4 for Pierre.

Many people, on seeing this solution, object, saying that the first two possible endings (H H and H T) are in reality the same one. They argue that if the fourth throw gives a head, then at that point, Blaise has his three heads and has won, so there would be no fifth throw. Accordingly, they argue, the correct way to think about the end of the game is that there are actually only three possibilities, namely

H   T H   T T

in which case, Blaise has a 2-to-1 advantage and the pot should be divided 2/3 for Blaise and 1/3 for Pierre, not 3/4 and

1/4. This reasoning is incorrect, but it took Pascal and Fermat some time to resolve this issue. Their colleagues, whom they consulted as they wrestled with the matter, had differing opinions. So if you are one of those people who finds this alternative argument appealing (or even compelling), take heart; you are in good company (though still wrong).

The issue behind the dilemma here is complex and lies at the heart of probability theory. The question is, What is the right way to think about the future (more accurately, the range of possible futures) and model it mathematically?

Interestingly, as far as we know, neither Pascal, Fermat, nor anyone else sought to resolve the issue empirically.* If you were actually to play out the completion of the game many times—that is, imagine the game had been halted after three tosses, with Blaise ahead 2 to 1, and then toss actual coins to complete the game—you would find that Blaise wins roughly 3/4 of the time. This would not constitute a mathematical proof, but it would indicate which solution is the right one. Many people today, when faced with such a puzzle about probabilities, do resort to a simulation (either in real life or on a computer) to help clarify their thoughts.

From today's perspective, we can explain exactly where the difficulty lies. You can indeed, if you wish, take the game's possible endings to be these:

H   T H   T T

---

*I'll come back to this interesting point later.

But if you do, you have to account for the frequencies of occurrence of each case. They are not all the same. If you do the math correctly, you find that the first outcome, H, occurs twice as often as either of the other two; the relative frequencies are 2 to 1 to 1, respectively. When you take account of those relative frequencies, you arrive at the same answer as you get with the previous approach: Blaise wins three out of four times.

Still confused? So was Pascal when he tried to understand the explanation of that very issue that Fermat had laid out in his previous letter. Let's take a closer look at how Pascal starts the second section of his letter:

> 2. *This is the method of procedure when there are two players. If two players, playing in several throws, find themselves in such a state that the first lacks two points and the second three of gaining the stake, you say it is necessary to see in how many points the game will be absolutely decided.*
>
> *It is convenient to suppose that this will be in four points, from which you conclude that it is necessary to see how many ways the four points may be distributed between the two players and to see how many combinations there are to make the first win and how many to make the second win, and to divide the stake according to that proportion. I could scarcely understand this reasoning if I had not known it myself before; but you also have written it in your discussion.*

As I observed earlier in this chapter, Cardano had already realized that the key was to look at the number of points each player would need in order to win, not the points they had already accumulated. In the second section of his letter to Fermat, Pascal acknowledged the tricky point we just encountered ourselves, that you have to look at all possible ways the game could have played out, ignoring the fact that the players would normally stop once one person had clearly won. But Pascal's words make clear that he found this hard to grasp, and he accepted it only because the great Fermat had explained it in his previous letter.

When Pascal continued his letter, it was to make the same simplifying assumption I did in formulating my version of the problem, reducing it from regular dice, which have six possible outcomes, to dice that have just two values, making them equivalent to tossing coins. When you toss four coins, there are $2 \times 2 \times 2 \times 2 = 16$ possible outcomes. Pascal enumerated them all in a table. The table has sixteen columns, each column listing the outcomes of the four throws. Beneath each column of the table he wrote a "1" if that outcome gives a win for player 1 and a "2" if it gives a win for player 2. Here is how he described this in his letter:

> *Then to see how many ways four points may be distributed between two players, it is necessary to imagine that they play with dice with two faces (since there are but two players), as heads and tails, and that they throw*

four of these dice (because they play in four throws). Now it is necessary to see how many ways these dice may fall. That is easy to calculate. There can be sixteen, which is the second power of four; that is to say, the square. Now imagine that one of the faces is marked a, *favorable to the first player. And suppose the other is marked* b, *favorable to the second. Then these four dice can fall according to one of these sixteen arrangements.*

$$a\ a\ a\ a\ a\ a\ a\ a\ b\ b\ b\ b\ b\ b\ b\ b$$
$$a\ a\ a\ a\ b\ b\ b\ b\ a\ a\ a\ a\ b\ b\ b\ b$$
$$a\ a\ b\ b\ a\ a\ b\ b\ a\ a\ b\ b\ a\ a\ b\ b$$
$$a\ b\ a\ b\ a\ b\ a\ b\ a\ b\ a\ b\ a\ b\ a\ b$$
$$1\ 1\ 1\ 1\ 1\ 1\ 2\ 1\ 1\ 1\ 2\ 1\ 2\ 2\ 2$$

and, because the first player lacks two points, all the arrangements that have two a's *make him win. There are therefore 11 of these for him. And because the second lacks three points, all the arrangements that have three* b's *make him win. There are 5 of these. Therefore it is necessary that they divide the wager as 11 is to 5.*

There is your method, when there are two players, whereupon you say that if there are more players, it will not be difficult to make the division by this method.

Having written out the table and noted which plays lead to a win for which player, Pascal had only to count the number of ways each can win—that is, the number of 1's and the number of 2's in his bottom row. His conclusion: for the game they were considering, the stake should be divided 11 to 5.

So far so good. But Pascal was by no means satisfied.

CHAPTER 3

On the Shoulders
of a Giant

*3. On this point, Monsieur, I tell you that this division for
the two players founded on combinations is very equitable and good, but that if there are more than two players, it is not always just and I shall tell you the reason for
this difference. I communicated your method to [some of]
our gentlemen, on which M. de Roberval made me this
objection:*

*That it is wrong to base the method of division on the
supposition that they are playing in* four *throws seeing
that when one lacks* two *points and the other* three, *there
is no necessity that they play* four *throws since it may
happen that they play but* two *or* three, *or in truth perhaps* four.

*Since he does not see why one should pretend to make
a just division on the assumed condition that one plays
four throws, in view of the fact that the natural terms of
the game are that they do not throw the dice after one of
the players has won; and that at least if this is not false, it*

should he proved. Consequently he suspects that we have committed a paralogism. *

As Pascal makes clear at the beginning of the third section of his letter to Fermat, others who have seen Fermat's proposed solution are also having trouble understanding why he ignored the fact that the two players would surely stop as soon as they recognized that one of them had already won.

NUMBER WAS THE KEY

When a single document, such as the August 24, 1654, letter from Pascal to Fermat, turns out to have a pivotal effect, the document itself can tell only part of the story. Prior circumstances must have prepared the ground. As Isaac Newton once wrote to the British physicist Robert Hooke, "If I have seen further [than certain other men] it is by standing upon the shoulders of giants." ** What Newton did not say is that getting up onto those shoulders often involves clambering over many other figures who have labored out of the limelight, making many small steps that together yield a ramp.

---

*The word *paralogism* is rarely used nowadays. It means a fallacious or illogical argument or conclusion. It comes from the late Latin *paralogismus,* from the Greek *paralogismos,* itself from *paralogos,* meaning "unreasonable" or "beyond (= *para*) logic."

**He was referring to his dependency on Galileo's and Kepler's work in physics and astronomy.

The development of probability took place during an extraordinary time in human history. The seventeenth century saw the birth not only of calculus and probability theory but also of modern science. These were not coincidences; all were part of a major shift in the way humans understand our world. Key to everything was number.

Though numbers themselves were first introduced around ten thousand years ago in Sumeria, their use was restricted to those who could both master cumbersome notations and become skilled in using the mechanical devices (such as the abacus) for carrying out computations. The modern, so-called Hindu-Arabic number system, developed in India between 200 and 700 A.D., was the first truly efficient way to record and compute with numbers, though this system was not available in the West until Leonardo of Pisa (whom later historians dubbed "Fibonacci") learned it from North African traders and described it in his book *Liber abaci (The Book of Calculating)*, which first appeared in 1202.

The new number system made it possible for anyone to master and use basic arithmetic. The first group to take advantage of this powerful new tool was the Italian merchants, for whom Leonardo primarily wrote his book. (His hometown, Pisa, was the European capital of what then constituted global trade.) But as a mathematician, Leonardo was interested in the theoretical aspects of the newly learned number system as well as its use, and as a result, his book also provided a source for scholars to study the new methods and to make use of this powerful new tool in their own researches.

By the time Luca Pacioli was born in 1445 in Sansepolcro, a small town more or less in the center of Italy, the Hindu-Arabic number system was widely known and used by both businesspeople and scholars. Like Leonardo before him, Pacioli was an excellent mathematician who is largely remembered not so much for his original contributions as for his book *Summa de arithmetica, geometrica, proportioni et proportionalita* (which I mentioned earlier), which cataloged, in a systematic way, all of what was known at the time. Published in Venice in 1494, the book draws heavily on both *Elements* and *Liber abaci* but also contains much that had been discovered since those two earlier great works appeared, particularly in algebra, where great strides had been made in the solution of polynomial equations.

Pacioli became a good friend of Leonardo da Vinci, whose own interests in mathematics and science are well known. It was Leonardo who drew the illustrations for Pacioli's book *Divina proportione (The Divine Proportion)*. Published in 1509, it is a study of what a nineteenth-century writer would rename the golden ratio, a mathematical constant (approximately equal to 1.618) first mentioned in *Elements*. This association is almost certainly the origin of the belief, which lacks the slightest supporting evidence, that Leonardo based many of his art works, including the *Mona*

Lisa, on that particular number. This belief is almost certainly false, but that has not prevented it from achieving the status of an urban legend in the world of art and, in due course, in popular culture.*

But I digress. Pacioli's *Summa* is significant in our story on two counts, first because it was so comprehensive. Occupying some six hundred densely printed folio pages, it treats arithmetic from both a theoretical and a practical standpoint, contains multiplication tables up to 60 × 60, provides a table of moneys, discusses weights and measures used in the various Italian states, and even provides one of the earliest accounts of double-entry bookkeeping, the technique that is so essential to all modern business and commerce. The second significance of *Summa* is that it includes the problem of the points—accompanied by Pacioli's incorrect solution—and thus provided a solid platform for the individual who set the stage for the Pascal-Fermat correspondence: Girolamo Cardano.

---

*Two other beliefs about this particular number are often mentioned in magazines and books: that the ancient Greeks believed it was the proportion of the rectangle the eye finds most pleasing and that they accordingly incorporated the rectangle in many of their buildings, including the famous Parthenon. These two equally persistent beliefs are likewise assuredly false and, in any case, are completely without any evidence. For one thing, tests have shown that human beings who claim to have a preference at all vary in the rectangle they find most pleasing, both from person to person and often the same person in different circumstances. Also, since the golden ratio is actually *not* a ratio of two whole numbers, it is impossible to construct (by measurement) a rectangle having that proportion, even in theory.

THE REMARKABLE MAN FROM MILAN

Fazio Cardano, Girolamo's father, was a successful lawyer based in Milan. He was also an accomplished mathematician who lectured on geometry both at the University of Pavia and at the Piatti Foundation in Milan. Leonardo da Vinci consulted him on questions of geometry. When Fazio was in his fifties, he had an affair with Chiara Micheria, a young widow in her thirties who was struggling to raise her three children. Chiara became pregnant, but before she was due to give birth, the plague hit Milan, and Fazio persuaded her to leave the city to have her child in the relative safety of nearby Pavia, where she stayed with wealthy friends of his. There, on September 24, 1501, she gave birth, naming her son Girolamo Cardano after his father. Her joy was short-lived, however. Shortly after the birth, she learned that her first three children, whom she had left behind in Milan, had all died of the plague.

Fazio and Chiara lived apart for many years but eventually married. When Girolamo grew up, he became his father's assistant. Fazio taught his son mathematics, leading the young man to contemplate an academic career. Fazio, however, wanted his son to study law, and the two at first argued until the father relented and allowed Girolamo to enter Pavia University (where he himself had studied) to read medicine.

When war broke out, the university was forced to close, and the young Cardano transferred to the University of

Padua to complete his studies. There he campaigned to become rector of the university. But for all his brilliance as a student, he was an outspoken and somewhat obnoxious individual who was not well liked. In the end, he beat his rival by only a single vote.

Cardano was one of the first major figures in mathematics to write an autobiography *(De vita propria liber [The Book of My Life])*, and as a result, we know far more about him than about most of his predecessors. Some might say he tells us far more than we care to know, since he was not sparing in the personal details.

Of his appearance, we know by his own words that he was skinny, had a long neck, a heavy lower lip, a wart over one eye, and a voice so loud that even his friends complained about it. We know, too, that he was constantly in ill health, suffering from, among other ailments, diarrhea, ruptures, kidney trouble, and palpitations. He was by his own account "ever hot-tempered, single-minded, and given to women." Elsewhere, he describes himself as "cunning, crafty, sarcastic, diligent, impertinent, sad, treacherous, magician and sorcerer, miserable, hateful, lascivious, obscene, lying," and "obsequious."

Of the characteristics that almost lost him the campaign for rector, he writes:

> This I recognize as unique and outstanding amongst my faults—the habit, which I persist in, of preferring to say above all things what I know to be displeasing to the ears of

my hearers. I am aware of this, yet I keep it up willfully, in no way ignorant of how many enemies it makes for me.

None of those character flaws seems to have prevented Cardano from giving the world its first systematic understanding of how to compute probabilities. Given his intellectual curiosity and abilities, his primary interest in the mathematics of games of chance may well have been scientific, but it is hard to discount another (perhaps even primary) motive. Throughout his life, Cardano was a compulsive gambler who needed every bit of help he could find at the gaming tables, from mathematics or any other source. (And he did find other sources of help. Once, when he suspected he was being cheated at cards, he took out the knife he always carried with him and slashed his opponent's face.)

Cardano's gambling began at an early age. Shortly after he moved to Padua, his father died, leaving him a small inheritance. It did not take Girolamo long to squander it all, and he turned to the gaming tables to maintain his lifestyle. From then on, he was addicted. "I have played not off and on but, as I am ashamed to say, every day," he wrote. Though his mathematical ability often helped him get the better of his opponents, things did not always go his way, and like any gambler, he lost more money than he won, sometimes causing significant personal hardship for him and those close to him.

Despite the gambling, Cardano's studies went well enough that in 1525, he was awarded his doctorate in medicine. He applied to join the College of Physicians in Milan, where his mother still lived. But by then the college was well aware of his difficult nature, and in spite of his exceptional performance as a student, college officials were reluctant to admit him. When they learned of his illegitimate birth, they had the excuse they wanted to reject his application.

On the advice of a friend, Cardano moved to Sacco, a small village about ten miles from Padua. Despite not being a member of the College of Physicians, he was able to set up a small, not very successful medical practice. It was there that he met Lucia Bandarini, whom he married in 1531. Since his small medical practice in Sacco did not provide enough income to support a wife, the couple moved to Gallarate, near Milan, the following year.

He applied once more to the College of Physicians in Milan, but again without success. Unable to practice medicine and losing money steadily at the gaming tables, he was eventually forced to pawn his wife's jewelry and some of his furniture. When a move into Milan itself brought no improvement, the couple had to enter the poorhouse.

Cardano's situation changed a little when he managed to secure Fazio's former post of lecturer in mathematics at the Piatti Foundation. When not occupied with teaching, he supplemented his salary by treating some patients. Strictly speaking, since he was not a member of the College of Physicians, this was not allowed, but he achieved some remarkable cures and his reputation as a doctor grew rapidly. With a client list that soon included wealthy people of influence in Milan—including some members of the college—it was surely only a matter of time before the college would be forced to admit him. But then, in 1536, still fuming at his continuing exclusion, he killed his chances by publishing a book attacking not only the college members' medical ability but their character as well:

> The things which give most reputation to a physician nowadays are his manners, servants, carriage, clothes, smartness and caginess, all displayed in a sort of artificial and insipid way.*

Not surprisingly, Cardano's application to join the college the following year was again rejected. Two years later, however, after the heat caused by his book had died down, the continued pressure from his many supporters finally persuaded the college to relent. They modified the rule regarding legitimate birth and admitted Cardano into their ranks.

In the same year, Cardano's first two mathematical books were published, beginning a prolific literary career that saw him writing on mathematics, medicine, philosophy, astronomy, and theology. In 1540, he resigned his mathematics post at the Piatti Foundation and spent the next two years doing nothing but gamble. From 1543 until 1552, he lectured on medicine at the universities of Milan and Pavia.

In 1545, he published his greatest mathematical work, *Ars magna,* a hugely influential book on algebra that pre-

---

*Girolamo Cardano, *Autobiography* (New York, 1930).

sented, for the first time, a method to solve a cubic equation. The result became known as Cardano's formula, even though he clearly stated that the method was discovered by Scipione del Ferro around 1500 and independently rediscovered by Niccolò Tartaglia in 1535. Cardano's publication of the method led to a heated dispute with Tartaglia, who had shown his formula to Cardano around 1539 on the strict condition that it be kept secret. Cardano's response was that he published the method only after he learned that del Ferro had obtained it much earlier, making his agreement with Tartaglia no longer binding.

In between his mathematical work, Cardano also found the time to engage in some engineering projects. His name was attached to a number of inventions, among them Cardano's suspension, the Cardano joint, and the Cardano shaft.

Cardano's wife, Lucia, died in 1546. By then, he was enjoying his reputation as the greatest physician in the world and basking in the fame his books and inventions had brought him. He became rector of the College of Physicians and later was appointed professor of medicine at Pavia University. With so many wealthy patients, he soon became rich.

But then disaster struck. His older son, Giambatista, who had qualified as a doctor in 1557, secretly married a young woman called Brandonia di Seroni. Cardano described her in his autobiography as "a worthless, shameless woman," a description that is richly supported by the evidence. With Cardano's financial support, though not his blessing, the young couple moved in with Brandonia's parents. However, the di

Seronis were interested only in what they could extort from Giambatista and his wealthy father, and Brandonia publicly mocked her husband for not being the father of their three children. These taunts eventually drove Giambatista to poison his wife, and after his arrest, he confessed to the crime. Giambatista's father hired the best lawyers, but at the trial, the judge decreed that to save his son's life, Cardano must come to terms with the di Seronis. They demanded a sum which Cardano could never have raised, and his son's fate was sealed. Giambatista was tortured in jail, his left hand was cut off, and on April 13, 1560, he was executed.

Cardano never forgave himself for failing to save his son. Moreover, as the father of a convicted murderer, his reputation was damaged beyond repair. Realizing he had to move, he secured a professorship of medicine at Bologna.

His time in Bologna was full of controversy. His reputation and his arrogant manner combined to create many enemies, and at one point, he humiliated a fellow medical professor in front of his students by pointing out errors in his lectures. After a few years, his colleagues tried to get the senate to dismiss him. They began spreading rumors that his lectures attracted few students.

Cardano had further problems with his children. His remaining son, Aldo, was a gambler and associated with individuals of dubious character. In 1569, the young man gambled away all his own clothes and other possessions in addition to a considerable sum of his father's money. In an attempt to pay off the debt, Aldo broke into his father's house and stole a large amount of cash and jewelry. Cardano sadly reported his son to the authorities, and Aldo was banished from Bologna.

In 1570, in what was probably a deliberate attempt to gain notoriety by offending the Church, Cardano cast the horoscope of Jesus Christ and wrote a book in praise of Nero, tormentor of Christian martyrs. Cardano was jailed on charges of heresy, but because of his fame, he was treated leniently and spent just a few months in prison. On his release, however, he was forbidden to hold a university post and was barred from further publication of his work. For a man such as Cardano, this could have been a more painful punishment than imprisonment, but once free, he went to Rome, and there he received an unexpectedly warm reception. He was granted immediate membership to the College of Physicians, and the pope, who had now apparently forgiven him, granted him a pension. It was in this period that he wrote his autobiography, although it was not published until 1643, long after his death. Cardano is said to have correctly predicted the exact date of his own death, September 21, 1576, but some historians have surmised that he achieved this triumph by committing suicide.

BOOK OF GAMES OF CHANCE

Cardano's *Liber de ludo aleae* (*Book of Games of Chance*) was published in 1663, long after his death, but according to his autobiography, he completed it in 1525, while still a young man, and rewrote it in 1565.* It is the first scientific study of dice rolling, based on the premise that there are fundamental principles governing the likelihood of particular outcomes. The book is partly observational (he had a lot of opportunity for observation) and partly a theoretical analysis of how chance events, such as particular outcomes of rolls of dice, aggregate when repeated many times. In modern parlance, it was the first study of frequentist probability.

Cardano did not use the word *probability*; rather, he talked of "chances." The word *probability*, which came later, derives from the Latin *probare* (to prove or test) and *ilis* (to be able) and may thus be understood as "able to be verified," where the verification is empirical—though it has also been suggested that "able to be believed" is a closer approximation to the original meaning.**

Not only did he not use the word *probability*, but Cardano did not even conceive of his work in a way that we would now classify as frequentist probability (i.e., counting relative frequencies). Rather, he saw his enterprise as very much "predicting the future" in the sense of formulating practical rules that would increase the likelihood of winning bets. (A more accurate description would be that the rules decrease the likelihood of losing. Cardano clearly realized

---

*The word *aleae* refers to dice games. *Aleatorius*, from the same Latin root, refers to games of chance in general.

**Attempts to trace origins and interpret earlier meanings of words are often fraught with difficulty, and in this case particularly so, since our present-day conception of probability, which has the two aspects of cataloging the past and predicting the future, was not developed until much more recently.

this himself, for he wrote, "The greatest advantage from gambling comes from not playing at all." A present-day echo of that observation can be found in the gambler's joke "I hope I break even tonight; I need the money.") The book is in part a how-to book for the gambler, peppered with asides about Cardano's own gambling experiences and beliefs.

But it is also part mathematics. Cardano defined, for the first time, what we now call *the probability of an event* as a fraction: the number of ways the event can occur divided by the total number of possible outcomes.* (He referred to the latter as the "circuit.") For example, if the event is getting an even number when a die is rolled (one-half the number of faces), the probability is 3/6, that is, 1/2. Here is how Cardano expressed this observation in his book:

> One-half the total number of faces always represents equality [of chance]; thus the chances are equal that a given point will turn up in three throws for the total circuit is completed in six, or again that one of three given points will turn up in one throw. For example, I can as easily throw one, three or five as two, four or six. The wagers there are laid in accordance with this equality if the die is honest.

He likewise observed that the probability of getting, say, a 3 or a 5 on a single roll is 2/6, or 1/3.

---

*For ease of understanding, I shall use the present-day term *probability* to describe Cardano's observations.

More penetrating, he observed that to obtain the probability of getting a certain outcome on two successive throws, you square the probability of getting it on a single throw. For example, the probability of getting a 6 twice in succession is $1/6 \times 1/6 = 1/36$. Similarly, the probability of getting an even number twice in succession is $1/2 \times 1/2 = 1/4$. Extending the reasoning, the probability of getting a 6 three times in a row will be $1/6 \times 1/6 \times 1/6 = 1/216$, and the probability of rolling three even numbers is $1/2 \times 1/2 \times 1/2 = 1/8$.

The more general version of this is that if an action occurs twice, and if the probability of an event or outcome E occurring the first time is p_E and the probability of an event or outcome F occurring the second time is p_F, then the probability that both will occur (in that order) is $p_E \times p_F$. This assumes that the first action does not influence the next—in modern parlance, the two occurrences of the action have to be *independent*. (Things will work out differently—in fact, you may not be able to calculate an answer—if the action does not satisfy this requirement.) For example, the probability of rolling a 6 on the first throw and any even number on the second is $1/6 \times 1/2 = 1/12$.

(If you remove the restriction on the order in which the 6 and an even number occur, there are more possibilities and, accordingly, the probability is bigger. The easiest way to calculate the probability in this case is to do what Fermat did in considering the problem of the points, and enumerate all the favorable outcomes. They are 2–6, 6–2, 4–6, 6–4, 6–6. Since there are 36 possible outcomes altogether, this gives the answer 5/36.)

More tricky is Cardano's next result: he calculated the probability of throwing a 1 or a 2 with not one die but a pair of dice. The probability of throwing a 1 or a 2 with a single die is 1/3, so the naive intuition is that with two dice, you double your chances—that is, to a probability of 2/3. But as Cardano observed, this is incorrect. The problem is that a 1 or a 2 could come up on both throws, and by adding the two individual probabilities, you would be counting this possibility twice. To allow for this, you have to subtract from the figure 2/3 the 1/9 probability of getting a 1 or a 2 on both throws. Hence, the correct answer is 2/3 − 1/9 = 5/9.

(Using Fermat's enumeration method, if you list all the thirty-six possible outcomes, you find that twenty of them include 1 or 2 at least once. Still another way to calculate the answer—more efficient when the number of possibilities grows very large—is to compute the probability that the event does *not* occur, since this avoids the problem of double counting. The probability of not rolling a 1 or a 2 with a single die is 4/6, or 2/3, so the probability of not rolling a 1 or a 2 with two dice is 2/3 × 2/3 = 4/9. Hence, since the probability of rolling *something* is 1, the probability of rolling a 1 or a 2 must be 1 − 4/9 = 5/9.)

With Cardano's analysis, the stage was set for Pascal and Fermat to make their breakthrough.

CHAPTER 4

A Man of Slight Build

. . . I replied to him [M. de Roberval] that I did not found my reasoning so much on this method of combinations, which in truth is not in place on this occasion, as on my universal method from which nothing escapes and which carries its proof with itself. This finds precisely the same division as does the method of combinations. Furthermore, I showed him the truth of the divisions between two players by combinations in this way. Is it not true that if two gamblers finding according to the conditions of the hypothesis that one lacks two points and the other three, *mutually agree that they shall play four complete plays, that is to say, that they shall throw four two-faced dice all at once,—is it not true, I say, that if they are prevented from playing the four throws, the division should be as we have said according to the combinations favorable to each? He agreed with this and this is indeed proved. But he denied that the same thing follows when they are not obliged to play the four throws.*

Advances in science are often delayed because the common wisdom holds that something is impossible. From the time of the ancient Greeks, it was believed that the future was in the hands of the gods—a matter of pure fate. Quantifying the way a pair of dice may fall, as Cardano had done, was one thing; predicting what future throws of the dice might bring (as in the unfinished game) was quite another.

As recently as 1756, over a hundred years after Pascal and Fermat had their famous exchange, the great French mathematician Abraham de Moivre (whom we shall meet later), the man who discovered the normal distribution that forms the basis of contemporary predictive statistics, firmly believed the future was determined by God. He wrote (in his book *Doctrine of Chances*):

Again, as it is thus demonstrable that there are, in the constitution of things, certain Laws according to which Events happen, it is no less evident from Observation, that those Laws serve to wise, useful and beneficent purposes to preserve the steadfast Order of the Universe, to propagate the several Species of Beings, and furnish to the sentient Kind such degrees of happiness as are suited to their State. But such Laws, as well as the original Design and Purpose of their Establishment, must all be from *without;* the *Inertia* of matter, and the nature of all created Beings, rendering it impossible that any thing should modify its own essence, or give to itself, or to anything else, an original determination or propensity. And hence, if we blind not ourselves with metaphysical dust, we shall be led, by a short and obvious way, to the acknowledgement of the great MAKER and GOVERNOR of all; *Himself all-wise, all-powerful* and *good.* °

Today, we have no difficulty in accepting that an event of pure chance may have sufficient "predictability" that we may calculate precisely the likelihoods pertaining to its outcome without needing to assume divine determination—the result really is a matter of pure chance. Indeed, today's religious person who sees the hand of God in all things simply accepts that the laws of probability are themselves a manifestation of God's ways. In short, we have taken the notions of undetermined random events and mathematically computed probabilities and folded them into our worldview. Quantum theory starts from the assumption that this is the nature of the very fabric of our universe.

But in the early seventeenth century, things looked very different. Pascal and Fermat not only had to figure out how to perform the calculation that resolves the problem of the unfinished game, but also had to do so within a worldview that considered what they were doing impossible. Prior to 1654, tomorrow was viewed as a matter of Fate, something over which a person had no control. In solving the problem, they were instrumental in changing that view, though as the passage from de Moivre indicates, it took a long time for the true import of their work to break through.

° A. de Moivre, *The Doctrine of Chances* (London: Pearson, 1718).

With the solution to the problem of the points and the subsequent acceptance of probability theory, mortals did have (a measure of) control over the future. An individual might not know with certainty what would happen, but by calculating the probabilities of various likely events, one might choose one's actions so as to minimize dangers and maximize preferred outcomes, or one might purchase insurance to facilitate recovery if things went badly. In short, Pascal and Fermat gave us the ability to manage risk.

An Abundance of Talents

Unlike Cardano, Pascal did not write an autobiography, and accordingly, we must rely on the scholarship of others to discover what kind of a man he was. In *Blaise Pascal: Mathematician, Physicist and Thinker About God,* biographer Donald Adamson writes that his subject was

> a man of slight build with a loud voice and somewhat overbearing manner . . . he lived most of his adult life in great pain. He had always been in delicate health, suffering even in his youth from migraine . . . [and was] precocious, stubbornly persevering, a perfectionist, pugnacious to the point of bullying ruthlessness yet seeking to be meek and humble.*

---

*D. Adamson, *Blaise Pascal: Mathematician, Physicist and Thinker About God* (Basingstoke: Palgrave Macmillan, 1995).

Historian René Taton, in his biography of Pascal in the *Dictionary of Scientific Biography,* provides this assessment of Pascal's life:

> At once a physicist, a mathematician, an eloquent publicist in the Provinciales . . . Pascal was embarrassed by the very abundance of his talents. It has been suggested that it was his too concrete turn of mind that prevented his discovering the infinitesimal calculus, and in some of the Provinciales the mysterious relations of human beings with God are treated as if they were a geometrical problem. But these considerations are far outweighed by the profit that he drew from the multiplicity of his gifts, [and] his religious writings are rigorous because of his scientific training.*

In December 1639, six years before Pascal had the conversation with the Chevalier de Méré that led to the famous correspondence with Fermat, the Pascal family left Paris to live in Rouen. Blaise's father, Étienne, had been appointed a tax collector for Upper Normandy. In February the following year, the young Blaise published his first genuine mathematical paper, *Essay on Conic Sections.* (The earlier "paper" he had presented to Mersenne's Academy in Paris as a teenager was a single-page affair—little more than a student

---

*René Taton, *Dictionary of Scientific Biography* (New York: Scribner, 1970–1990).

exercise, though Blaise's young age at the time made it a portent of greater things to come.)

It was in Rouen, from 1642 to 1645, that he worked on the Pascaline, his mechanical calculator. The only previous attempt to construct such a device was in 1624 by a man called Schickard, so Pascal's machine was the second mechanical calculator ever built.

The year after Blaise finished work on his device, his father slipped on the ice and broke his hip. Two young men were hired to care for Étienne. The caretakers were brothers in a nearby religious community of Jansenists. Jansenism was a proselytizing Catholic sect that preached an extreme form of asceticism, sacrifice, and strict adherence to the Scriptures as the only path to salvation. Though the two brothers seemed to have little effect on Étienne Pascal, they were more successful with the young and impressionable son, who adopted their faith. From that point forward, there was to be no pleasurable social life for young Blaise. Intellectual pursuits such as mathematics or science were also distractions to be avoided.

For a short while, Pascal renounced mathematics as he fought to save his soul from eternal damnation. The terror of such a fate eventually became too great for the young man to bear, however, and he became ill, suffering severe headaches and a partial paralysis. His doctor advised him that for the sake of his health, he should abandon the Jansenist ways and lead a life more normal for a young man. Although he would remain strongly religious for the remainder of his all-too-short life, Blaise resumed normal activities. Indeed, he did so with vigor, adding regular visits to the gaming rooms to his earlier academic pursuits. It was at the gambling table that Pascal met the Chevalier de Méré, a keen gambler with sufficient mathematical ability to figure out for himself some of the more favorable odds.

Pascal's first focus, when he resumed his researches, was in physics. He carried out a series of experiments in which he observed that the pressure of the atmosphere decreases with height and so deduced that, at some height, it must thin out to a vacuum. In September 1647, he informed Descartes of his finding when the great philosopher came on a visit. Descartes refused to accept Pascal's reasoning, and the pair argued about it for two days. Reporting on the dispute afterward in a letter to the physicist Christiaan Huygens, Descartes noted rather haughtily that Pascal "has too much vacuum in his head." In October 1647, Pascal published his findings under the title *New Experiments Concerning Vacuums*, which led to disputes with a number of scientists who, like Descartes, did not believe in a vacuum.

Étienne died in September 1651. The event prompted Blaise to write to one of his sisters describing his Christian views on death and how they applied to their deceased father. The ideas he expressed in that letter were to form the basis for *Pensées*, Pascal's most famous work of philosophy, a collection of personal thoughts on human suffering and faith in God, which he began in late 1656 and continued to work on during 1657 and 1658. *Pensées* was finally published in

1670, eight years after Pascal's death. While it is a philo-sophical text, *Pensées* is not totally devoid of mathematical content. Among its essays is one that has come to be known as *Pascal's Wager*, an argument that belief in God is rational. Stripped of its numerical calculations, Pascal's case is essen-tially that if God does not exist, one will lose little by believ-ing in the supreme being, while if God does exist, one will lose everything by not believing. We'll look at this argument in its full mathematical glory in Chapter 7.

In 1653, Pascal wrote *Treatise on the Equilibrium of Liq-uids,* in which he explains what is nowadays known as Pas-cal's law of pressure. At the same time, he resumed his earlier investigations of conic sections and proved some im-portant theorems in projective geometry. In particular, he completed a work titled *The Generation of Conic Sections,* most of which he had written in 1648. This was meant to be the first part of a comprehensive treatise on conics, but he never got back to it.

His *Traité du triangle arithmétique (Treatise on the Arith-metical Triangle),* completed and printed in 1654 but not re-leased until 1665, contained the numerical triangle that today bears his name. Although he was not the first to study this particular mathematical structure, his insights were far more illuminating than anything written beforehand. Pascal's results concerning binomial coefficients were to lead Isaac Newton to his discovery of a hugely important result known as the general binomial theorem (the binomial theorem for fractional and negative powers).

The full title of Pascal's work, rarely given in historical accounts, was *Traité du triangle arithmétique, avec quelques autres petits traités sur la même matière.* The additional clause translates as "with some other small treatises on the same topic," and the essay covers, among other things, Pascal's fairly comprehensive treatment of everything then known about probability theory (as we now call it). He sent a copy to Fermat, who referred to it in one of his letters to Pascal.

It was around the time he started his correspondence with Fermat about the problem of the points, in the summer of 1654, that Pascal first fell ill. Although Pascal did not know it at the time, he was suffering from the first stages of the stomach cancer that would eventually kill him. In one of his letters, written in July 1654, he told Fermat, "though I am still bedridden, I must tell you that yesterday evening I was given your letter." Before his correspondence with Fermat began, Pascal had been unsure if his own initial attempt at a solution was correct, and he first discussed the matter with his colleague Pierre de Carcavi, another member of Mersenne's Academy. Carcavi in turn suggested that Pascal contact Fermat, acknowledged to be greatest mathematician alive.

A few weeks after writing the August 24 letter that is the focus of our story, Pascal almost lost his life in an accident. The horses pulling his carriage bolted, and the carriage was left hanging over a bridge above the river Seine. Although he was rescued without injury, the experience seems to have affected him psychologically, and not long afterward, he underwent another religious experience. On November 23, 1654, he pledged his life to Christianity once again and, soon after, made the first of several visits to the Jansenist monastery Port-Royal des Champs, about twenty miles southwest of Paris. He began to publish anonymous works on religious topics, including eighteen *Provincial Letters* published during 1656 and early 1657. These were written in defense of his friend Antoine Arnauld, an opponent of the Jesuits and a defender of Jansenism, who was on trial before the faculty of theology in Paris for his controversial religious works.

As the pain from the malignancy in his stomach increased, Pascal lost interest in science. In 1658, he tried to do some mathematics, much of it in correspondence with others, but that was largely as a distraction from the pain and he soon gave up. He spent his last years giving to the poor and going from church to church in Paris, attending one religious service after another.

On August 19, 1662, at age thirty-nine, Blaise Pascal died in Paris, in intense pain after the cancer reached his brain. He left behind a legacy of ideas made all the greater by his brief correspondence with Fermat in 1654.

FIRST LETTERS

The letter from Pascal to Fermat that began their collaboration has unfortunately been lost, and our knowledge of the two men's correspondence begins with Fermat's undated reply. After a brief discussion of the part of Pascal's letter that Fermat has found problematic—concerning the ever troublesome issue of how to handle games where one player wins in fewer than the maximum number of rounds—Fermat explains the difficulty he sees in Pascal's approach:

> *But you proposed in the last example in your letter (I quote your very terms) that if I undertake to find the six in eight throws and if I have thrown three times without getting it, and if my opponent proposes that I should not play the fourth time, and if he wishes me to be justly treated, it is proper that I have 125/1296 of the entire sum of our wagers.*
>
> *This, however, is not true by my theory. For in this case, the three first throws having gained nothing for the player who holds the die, the total sum thus remaining at stake, he who holds the die and who agrees to not play his fourth throw should take 1/6 as his reward.*
>
> *And if he has played four throws without finding the desired point and if they agree that he shall not play the fifth time, he will, nevertheless, have 1/6 of the total for his share. Since the whole sum stays in play it not only follows from the theory, but it is indeed common sense that each throw should be of equal value.*
>
> *I urge you therefore [to write me] that I may know whether we agree in the theory, as I believe [we do], or whether we differ only in its application.*
>
> I am, most heartily, etc.,
> Fermat

Pascal acknowledges the validity of Fermat's objection in his next letter, dated Wednesday, July 29, 1654, which begins:

> 1. *Impatience has seized me as well as it has you, and although I am still abed, I cannot refrain from telling you that I received your letter in regard to the problem of the points yesterday evening from the hands of M. Carcavi, and that I admire it more than I can tell you. I do not have the leisure to write at length, but, in a word, you have found the two divisions of the points and of the dice with perfect justice. I am thoroughly satisfied as I can no longer doubt that I was wrong, seeing the admirable accord in which I find myself with you.*

Pascal was a mathematician of formidable powers. But as their respective mathematical records make plain, Fermat was far superior, and anyone who reads the entire correspondence between the two will see readily that Fermat was the dominant collaborator.

PASCAL'S UNIVERSAL METHOD

Fermat's initial reply to Pascal and the famous August 24 letter from Pascal to Fermat make it clear that the two men had adopted different approaches to solving the problem of the points. While Fermat listed explicitly all the possible combinations of outcomes that could arise in the completion of the game, Pascal adopted a different strategy. Since both methods are correct, they will of course yield the same result, as Pascal made clear in his letter:

> *I did not found my reasoning so much on this method of combinations, which in truth is not in place on this occasion, as on my universal method from which nothing escapes and which carries its proof with itself. This finds precisely the same division as does the method of combinations.*

The "universal method" that Pascal refers to is an instance of what is generally known as a *recursive method*. Recursion is the process whereby a sequence of numbers is generated by a rule that gives the new number in terms of the last number produced (or sometimes the last two numbers). The most famous example of a recursion is the rule for generating the Fibonacci numbers. As an exercise in his book *Liber abaci*, Leonardo of Pisa (Fibonacci) posed a problem about the growth of a rabbit colony. To solve the problem, you have to generate the sequence of numbers that starts with two 1's and grows according to the rule that the new number at each stage is the sum of the previous two numbers. When you follow this rule, you get the sequence

$$1, 1, 2, 3, 5, 8, 13, 21, 34, 55, 89, \ldots$$

Pascal approaches the problem of the points by looking at the quantity $e(a, b)$ that gives the share of the stake that the first player should be given if player 1 requires a winning throws to win and player 2 requires b winning throws. Clearly, if the numbers a and b are equal, then $e(a, b) = 1/2$. The idea is to see how $e(a, b)$ changes when each player wins one more throw. This leads to an algebraic expression for $e(a, b)$ in terms of $e(a - 1, b)$ and $e(a, b - 1)$, and you can solve the problem of the points by using recursion to calculate $e(2, 3)$, the desired share in the particular game they considered.

The main drawback to Pascal's approach was that it requires some complicated algebra dependent upon the theory of combinations he worked out in connection with his famous triangle. Nevertheless, when carried out correctly (not an easy matter), it does lead to the correct result, and the same approach was subsequently used by Abraham de Moivre, Joseph Louis Lagrange, Pierre-Simon Laplace, and other mathematicians to find general methods for solving a variety of other problems.

A detailed description of Pascal's recursive solution to the problem of the points quickly becomes too technical for this book. But my main reason for not including it is that Fermat's solution is simply much better.

Outside observers often assume that the more complicated a piece of mathematics is, the more mathematicians admire it. Nothing could be further from the truth. Mathematicians admire elegance and simplicity above all else, and the ultimate goal in solving a problem is to find the method that does the job in the most efficient manner. Though the major accolades are given to the individual who solves a particular problem first, credit (and gratitude) always goes to those who subsequently find a simpler solution.

Pascal's solution to the problem of the points was far more complicated than Fermat's and required some sophisticated—and daunting!—algebra. It is hard to follow even for a professional mathematician. But for all its complicated appearance, following it (or even producing it in the first place) is, again from the perspective of the professional mathematician, routine. Fermat's approach is undoubtedly more pedestrian, requiring only that you list all the possibilities and then simply count them. But by penetrating to the heart of the problem and doing *just what is required* to get the answer, Fermat's approach shows true genius.

CHAPTER 5

The Great Amateur

. . . I therefore replied as follows:

It is not clear that the same gamblers, not being constrained to play the four throws, but wishing to quit the game before one of them has attained his score, can without loss or gain be obliged to play the whole four plays, and that this agreement in no way changes their condition? For if the first gains the two first points of four, will he who has won refuse to play two throws more, seeing that if he wins he will not win more and if he loses he will not win less? For the two points which the other wins are not sufficient for him since he lacks three, and there are not enough [points] in four throws for each to make the number which he lacks.

It certainly is convenient to consider that it is absolutely equal and indifferent to each whether they play in the natural way of the game, which is to finish as soon as one has his score, or whether they play the entire four throws. Therefore, since these two conditions are equal and indifferent, the division should be alike for each. But since it is just when they are obliged to play the four throws as I have shown, it is therefore just also in the other case.

That is the way I prove it, and, as you recollect, this proof is based on the equality of the two conditions true and assumed in regard to the two gamblers, the division is the same in each of the methods, and if one gains or loses by one method, he will gain or lose by the other, and the two will always have the same accounting.

Pascal convinces himself that it is correct to consider the imagined completion of the game in such a way that the players complete all remaining throws, not stopping when one player has won, the same way that many people today resolve the issue. It might be a waste of time to play after one person has won, they say, but there is nothing *wrong* with doing that. So it cannot lead to an incorrect answer if you look at it that way. Today, however, accustomed as we are to the fact that probability theory can predict the future, this argument is easier to accept than it was in Pascal and Fermat's time, when it was not even clear that you *could* predict the future, let alone *how* to do it.

THE LAWYER IN TOULOUSE

Fermat was born on August 17, 1601, in the French city of Beaumont-de-Lomagne, where his father was a wealthy leather merchant and second consul of the city. Pierre had one brother and two sisters and was almost certainly brought up in the town of his birth. Although there is little evidence concerning his early education, most likely it took place at the local Franciscan monastery.

Fermat attended the University of Toulouse before moving to Bordeaux in the second half of the 1620s. In Bordeaux, he began his first serious mathematical researches, producing important work on maxima and minima.

From Bordeaux, Fermat went to Orléans, where he studied law at the university. He received a degree in civil law and purchased the office of counselor at the parliament in Toulouse. By 1631, he was a lawyer and government official in Toulouse, which allowed him to change his name from Pierre Fermat to Pierre de Fermat, signifying minor nobility. For the remainder of his life, he lived in Toulouse but regularly spent time in his home town of Beaumont-de-Lomagne and the nearby town of Castres.

Fermat's career as a lawyer and jurist progressed rapidly. Initially, he worked in the lower chamber of the parliament, but in 1638, he was appointed to the higher chamber, and in 1652, he was promoted to the highest level at the criminal court. Still further promotions followed, though his rapid rise does not necessarily mean that he put any unusually great effort into his duties. Promotion was done mostly on seniority, and when the plague struck the region in the early 1650s, many of the older men died. Fermat himself survived the plague in 1653.

The common description of Fermat as "the great amateur" reflects only that he did not pursue mathematics for a living and did not publish his work. In reality, he devoted enormous time and effort to the subject, corresponding on a regular basis with some of the best mathematicians in Europe. He had little interest in practical problems, focusing most of his attention on number theory, the branch of mathematics that most mathematicians regard as the "purest of the pure." As recently as 1944, the great English mathematician G. H. Hardy remarked* that number theory (his preferred subject, also) had absolutely no practical use, a claim that remained true until the 1970s, when the theory of prime numbers was used to construct a highly secure encryption system that is now used to protect most of the confidential traffic sent over the Internet.

In his 1994 book *The Mathematical Career of Pierre de Fermat (1601–1665),* Michael Sean Mahoney describes Fermat as "secretive and taciturn, he did not like to talk about himself and was loath to reveal too much about his thinking."** Fermat's secrecy was legendary. His standard method for informing others of his work was to send out letters in which he stated his results, giving little or no indication as to how he had obtained them, let alone complete proofs or detailed solutions. Though occasionally one

---

*G. H. Hardy made the statement about number theory in his autobiographical book *A Mathematician's Apology*, canto ed. (Cambridge: Cambridge University Press, 2001).

**Michael Sean Mahoney, *The Mathematical Career of Pierre de Fermat (1601–1665)*, 2nd rev. ed. (Princeton, NJ: Princeton University Press, 1994), xii.

of his claims turned out to be wrong, the vast majority were correct, and accordingly, the receipt of a letter from Fermat amounted to a direct challenge: "I, Fermat, can do this; can you?"

Fermat's avoidance of publication suggests he had no ambition for fame. In a letter dated April 26, 1636, in which he described to Marin Mersenne, the founder of Mersenne's Academy, some results on geometric spirals, he wrote, "I will share all of this with you whenever you wish and do so without any ambition, from which I am more exempt and more distant than any man in the world." On the other hand, he did regularly inform others of his results, so it would appear he desired recognition from those he regarded as his peers—a decidedly elite group.

Others have interpreted his behavior differently. Jean-Baptiste Colbert, a leading figure in France at the time, wrote that "Fermat, a man of great erudition, has contact with men of learning everywhere. But he is rather preoccupied, he does not report cases well and is confused."

Certainly, his letters to Pascal about the problem of the unfinished game were far shorter—though mathematically much more insightful and to the point—than those written by Pascal, although the entire correspondence was written in exquisitely polite language, and each writer displayed great respect for the other.

This mutual respect was clearly genuine. In a letter to Pierre de Carcavi, Fermat wrote: "I am delighted to have had opinions conforming to those of M. Pascal, for I have infinite esteem for his genius."*

Fermat appears to have had less regard for Descartes. When asked by Mersenne for his opinion of Descartes' *La dioptrique,* Fermat dismissed it as "groping about in the shadows" and suggested that Descartes had not correctly deduced his law of refraction.

Descartes, needless to say, was livid. His anger grew even worse when he recognized that Fermat's work on maxima, minima, and tangents could be seen as reducing the importance of his own work, *La géométrie.* In retaliation, Descartes attacked Fermat's work. Roberval and Étienne Pascal became involved in the ensuing argument, and eventually so did another prominent colleague, Desargues, whom Descartes asked to act as a referee. In the end, Fermat proved correct, and eventually Descartes admitted this, albeit somewhat churlishly: "seeing the last method that you use for finding tangents to curved lines, I can reply to it in no other way than to say that it is very good and that, if you had explained it in this manner at the outset, I would have not contradicted it at all."

For Descartes, however, this was not the end of the matter. Although he wrote to Fermat praising Fermat's work on determining the tangent to a cycloid (which was correct), he wrote to Mersenne claiming that it was incorrect and that Fermat was a second-rate mathematician and thinker. Descartes' importance was such that his words severely

---

*Ibid., p. 61.

damaged Fermat's reputation, but there is no indication that this bothered Fermat one iota.

For eleven years, from 1643 to 1654, Fermat largely lost touch with his scientific colleagues in Paris. In part this was due to the pressure of work, which kept him from devoting so much time to mathematics. Second, the Fronde, a civil war waged in France, affected Toulouse in 1648. Finally, there was the plague of 1651, with its near-terminal consequence for Fermat himself.

Fermat's correspondence with the Paris mathematicians resumed in 1654, when Blaise Pascal wrote him to ask his advice on the problem of the unfinished game. After his collaboration with Pascal, Fermat lived another eleven years, remaining active in mathematics and continuing to work as he had always done, by sending letters to others, describing, without proofs, his latest results. He died at age sixty-three, on January 12, 1665, in Castres, France.

Though their four months of correspondence in 1654 would change the world, Fermat and Pascal never met in person. On one occasion, Fermat tried to arrange a meeting. Hearing that his friend was in Clermont, not too far from his home in Toulouse, he wrote suggesting they try to get together. Sadly, Pascal wrote back immediately to say that his health was now so bad he could not manage even so short a journey by carriage and would be returning soon to Paris by river. It had, he said, taken him twenty-one days to get to Clermont, as he was unable to travel more than a very short distance each day.

Two years later, Pascal was dead.

CHAPTER 6

Terrible
Confusions

4. Let us follow the same argument for three *players
and let us assume that the first lacks* one *point, the second*
two, *and the third* two. *To make the division, following
the same method of combinations, it is necessary to first
discover in how many points the game may be decided as
we did when there were two players. This will be in three
points for they cannot play three throws without necessarily arriving at a decision.*

*It is now necessary to see how many ways three
throws may be combined among three players and how
many are favorable to the first, how many to the second,
and how many to the third, and to follow this proportion
in distributing the wager as we did in the hypothesis of
the two gamblers.*

Although Pascal has solved the problem of the points to
his satisfaction using his recursive method, he still struggles

to understand Fermat's approach.* The root of his difficulty is, as we noted earlier, the vexing question of what difference it makes when, in the imagined continuation of the game, the players stop playing as soon as one of them has won. To try to grasp Fermat's solution, Pascal does what all mathematicians are trained to do: he makes the problem slightly more complicated, to see what happens. Suppose, he says, there are not two but three players, and they are rolling a die that has not two equally likely outcomes but three (say, two faces marked *a*, two marked *b*, and two marked *c*). He analyzes this modified game using Fermat's method, insofar as he understands it (and he makes it clear he is not at all sure that he does):

> *It is easy to see how many combinations there are in all. This is the third power of 3; that is to say, its cube, or 27. For if one throws three dice at a time (for it is necessary to throw three times), these dice having three faces each (since there are three players), one marked a favorable to the first, one marked b favorable to the second, and one marked c favorable to the third,—it is evident that these three dice thrown together can fall in 27 different ways as:*

---

*This chapter is a bit more technical than the remainder of the book, and many readers may prefer to skip through with merely a cursory glance at the details. There is no shame in that; after all, Pascal found it hard going as well. The chapter title refers to Pascal, not to my reader!

aaa	aaa	aaa	bbb	bbb	bbb	ccc	ccc	ccc
aaa	bbb	ccc	aaa	bbb	ccc	aaa	bbb	ccc
abc	abc	abc	abc	abc	abc	abc	abc	abc
111	111	111	111	1	1	111	1	1
	2			2	222	2		2
		3				3	3	3 333

By analogy with the way Fermat proceeded with the original game, Pascal charts out all the possible combinations of the three-die outcomes (of which there are $3 \times 3 \times 3 = 27$ in all); notes which player wins when each comes up (he numbers the players 1, 2, and 3); and then adds up the number of ways each can win. He sees that things are quite a bit more complicated this time around, since some die-rolls result in two players winning. (Pascal's three-person game is best three of seven, and the seven rolls can end with the score 3 to 3 to 1.) He continues:

> Since the first lacks but one point, then all the ways in which there is one a are favorable to him. There are 19 of these. The second lacks two points. Thus all the arrangements in which there are two b's are in his favor. There are 7 of them. The third lacks two points. Thus all the arrangements in which there are two c's are favorable to him. There are 7 of these. If we conclude from this that it is necessary to give each according to the proportion 19, 7, 7, we are making a serious mistake and I would hesitate to believe that you would do this. There are several cases favorable to both the first and the second, as abb has the a

which the first needs, and the two b's which the second needs. So too, the acc is favorable to the first and third.

By this reasoning, Pascal finds himself faced with outcomes that will give two players a win, which he handles by giving each half a point:

It therefore is not desirable to count the arrangements which are common to the two as being worth the whole wager to each, but only as being half a point. For if the arrangement acc occurs, the first and third will have the same right to the wager, each making their score. They should therefore divide the wager in half. If the arrangement aab occurs, the first alone wins. It is necessary to make this assumption.

With the possibility of using halves in dividing the pot, he continues his calculation:*

There are 13 arrangements which give the entire wager to the first, and 6 which give him half and 8 which are worth nothing to him. Therefore if the entire sum is one pistole, there are 13 arrangements which are each worth one pistole to him, there are 6 that are each worth ½ a pistole, and 8 that are worth nothing.

---

*He refers to the stake in units of a *pistole*. *Pistole* is the French name given to a Spanish gold coin used in the sixteenth century. The name was also used to refer to other European gold coins of roughly the same value as the Spanish coin.

Then in this case of division, it is necessary to multiply

	13	*by one pistole which makes*	13
	6	*by one half which makes*	3
	8	*by zero which makes*	0
Total	27	*Total*	16

and to divide the sum of the values 16 by the sum of the arrangements 27, which makes the fraction 16/27 and it is this amount which belongs to the first gambler in the event of a division; that is to say, 16 pistoles out of 27.

The shares of the second and the third gamblers will be the same:

There are	4	*arrangements which are worth 1 pistole; multiplying,*	4
There are	3	*arrangements which are worth 3/2 pistole; multiplying,*	1½
And	20	*arrangements which are worth nothing*	0
Total	27	*Total*	5½

Therefore 5 pistoles belong to the second player out of 27, and the same to the third. The sum of the 5½, 5½, and 16 makes 27.

This answer is wrong, and he knows it. He has arrived at an incorrect conclusion because he has not really grasped

Fermat's method. His certainty about the source of his difficulty becomes evident as he continues:

> 5. *It seems to me that this is the way in which it is necessary to make the division by combinations according to your method, unless you have something else on the subject which I do not know. But if I am not mistaken, this division is unjust.*

Pascal is still hung up on the issue of when, in practice, the players in the imagined continuation would stop playing because someone had won. He struggles to clarify his thoughts on the matter:

> *The reason is that we are making a false supposition,—that is, that they are playing three throws without exception, instead of the natural condition of this game which is that they shall not play except up to the time when one of the players has attained the number of points which he lacks, in which case the game ceases.*
>
> *It is not that it may not happen that they will play three times, but it may happen that they will play once or twice and not need to play again.*

He hopes that by comparing the original, two-person game with his three-person variation, he will be able to pin down the source of his difficulty in understanding Fermat's approach.

But, you will say, why is it possible to make the same assumption in this case as was made in the case of the two players? Here is the reason: In the true condition [of the game] between three players, only one can win, for by the terms of the game it will terminate when one [of the players] has won. But under the assumed conditions, two may attain the number of their points, since the first may gain the one point he lacks and one of the others may gain the two points which he lacks, since they will have played only three throws. When there are only two players, the assumed conditions and the true conditions concur to the advantage of both. It is this that makes the greatest difference between the assumed conditions and the true ones.

If the players, finding themselves in the state given in the hypothesis,—that is to say, if the first lacks one point, the second two, and the third two; and if they now mutually agree and concur in the stipulation that they will play three complete throws; and if he who makes the points which he lacks will take the entire sum if he is the only one who attains the points; or if two should attain them that they shall share equally, in this case, the division should be made as I give it here. The first shall have 16, the second 5½, and the third 5½ out of 27 pistoles, and this carries with it its own proof on the assumption of the above condition.

But if they play simply on the condition that they will not necessarily play three throws, but that they will only play until one of them shall have attained his points, and *that then the play shall cease without giving another the opportunity of reaching his score, then 17 pistoles should belong to the first, 5 to the second, and 5 to the third, out of 27. And this is found by my general method which also determines that, under the proceeding condition, the first should have 16, the second 5½, and the third without making use of combinations,—for this works in all cases and without any obstacle.*

FERMAT STUMBLES

When Fermat receives Pascal's long letter (there is a little more of it to come as we continue our story, but we now have the hard part behind us), he sees at once where his correspondent has gone wrong, and writes back immediately. Perhaps out of kindness to Pascal, or perhaps because, having solved the problem of the points to his own satisfaction, his interests have now moved on to other things, he buries his assessment of Pascal's confused analysis of the three-person game in the second section of his letter and starts instead with something that will make his correspondent feel good, namely, his work on the arithmetic triangle.

Saturday, August 29, 1654

Monsieur,

1. Our interchange of blows still continues, and I am well pleased that our thoughts are in such complete adjustment as it seems since they have taken the same di-

rection and followed the same road. Your recent Traité du triangle arithmétique *and its applications are an authentic proof and if my computations do me no wrong, your eleventh consequence went by post from Paris to Toulouse while my theorem, on figurate numbers, which is virtually the same, was going from Toulouse to Paris. I have not been on watch for failure while I have been at work on the problem and I am persuaded that the true way to escape failure is by concurring with you. But if I should say more, it would be of the nature of a Compliment and we have banished that enemy of sweet and easy conversation.*

It is now my turn to give you some of my numerical discoveries, but the end of the parliament augments my duties and I hope that out of your goodness you will allow me due and almost necessary respite.

Now Fermat cuts to the chase:

2. I will reply however to your question of the three players who play in two throws. When the first has one [point] and the others none, your first solution is the true one and the division of the wager should be 17, 5, and 5. The reason for this is self-evident and it always takes the same principle, the combinations making it clear that the first has 17 chances while each of the others has but five.

3. For the rest, there is nothing that I will not write you in the future with all frankness.

In other words, Pascal can solve the three-person game correctly when he uses his own (considerably more complicated) recursive method, but he still has not grasped Fermat's enumeration approach. Not for the first time has a mathematician—even one as accomplished as Pascal—found that it is possible to get the right solution by correctly applying an appropriate method while not really understanding the subtleties of the problem. [*]

At this point, again perhaps out of kindness to Pascal, Fermat changes the topic:

> *Meditate however, if you find it convenient, on this theorem: The squared powers of 2 augmented by unity are always prime numbers. [That is,]*
>> *The square of 2 augmented by unity makes 5 which is a prime number;*
>> *The square of the square makes 16 which, when unity is added makes 17, a prime number;*
>> *The square of 16 makes 256 which, when unity is added, makes 257, a prime number;*
>> *The square of 256 makes 65536 which, when unity is added, makes 65537, a prime number;*
>> *and so to infinity.*

---

[*] For example, few university undergraduates fully understand differential calculus, but that does not prevent them from using it to solve problems correctly. That was certainly true for me when I was a student. I came to understand the method only many years later, when I was faced with teaching it to students of my own. I gather that my experience is by no means unique among professors of mathematics.

This is a property whose truth I will answer to you. The proof of it is very difficult and I assure you that I have not yet been able to find it fully. I shall not set it for you to find unless I come to the end of it.

This theorem serves in the discovery of numbers which are in a given ratio to their aliquot parts, concerning which I have made many discoveries. We will talk of that another time.

I am, Monsieur, yours etc.

Fermat

At Toulouse, the twenty ninth of August, 1654

By a remarkable good fortune in terms of social equality, in the very letter in which Fermat corrects Pascal's faulty reasoning, he makes a major blunder of his own—one of the few in his entire career. His claim is that each of the numbers

$$2^{2^n} + 1$$

is prime. Nowadays, numbers of this form are called *Fermat numbers*, generally denoted by F_n.

Fermat lists the first few cases: $F_1 = 2^2 + 1 = 5$; $F_2 = 2^4 + 1 = 17$; $F_3 = 2^8 + 1 = 257$; $F_4 = 2^{16} + 1 = 65,537$, and observes, correctly, that each is prime. He then claims that all numbers of this form are prime. This is not so. The great Swiss mathematician Leonhard Euler discovered in 1732 that the very next Fermat number, $F_5 = 4,294,967,297$, is not prime, being the product of the primes 641 and 6,700,417.

In fact, no Fermat number beyond F_4 has been shown to be prime, and all Fermat numbers from F_5 through F_{32} have been shown, through the use of computers, to be composite. Some mathematicians have suggested that perhaps no Fermat number is prime other than the four he looked at. Even Fermat made a mistake occasionally.

CHAPTER 7

Out of the Gaming Room

6. These, Monsieur, are my reflections on this topic on which I have no advantage over you except that of having meditated on it longer, but this is of little [advantage to me] from your point of view since your first glance is more penetrating than are my prolonged endeavors.

Opening the final section of his letter, Pascal makes it clear that he fully realizes Fermat is by far the better mathematician. Although he himself solved the problem of the points, much of his long letter is devoted to his attempt to understand Fermat's clearly superior (because simpler and more insightful) method. He appreciates that whereas he labored long and hard to find a solution, Fermat almost certainly saw at once how to set about it. Such is the mark of a truly great mathematician, of which history has seen but a handful.

Neither man could possibly have appreciated the true significance of what they had done. Even the greatest intellects are prisoners of the age in which they live. Pascal and Fermat would never know that their exchange would give humanity a way to see into the future, changing life dramatically and forever.

ONLY THE GODS KNOW WHAT TOMORROW WILL BRING

What made the problem of the points such a challenge to Pascal (perhaps to Fermat as well)* and all who had wrestled with it earlier, was not that people did not know how to calculate odds. They did. They also recognized that the odds tell you something about the next roll of the die. The Chevalier de Méré himself, the man who first informed Pascal of the problem, was sufficiently skilled in calculating odds to do quite well at the gaming tables. A winning strategy he often used was to bet repeatedly on outcomes that have just a narrow margin in their favor. He knew, for instance, that the probability of rolling a 6 with one die rises above 1/2 when you throw four times. (The probability is 51.77 percent.) Thus, by betting a small amount on 6 repeatedly, he could guarantee himself a profit—at least provided he had sufficient wealth to stay in the game long enough for the odds to have their effect. Another strategy the Chevalier adopted was to bet that he could throw a double 6 in twenty-four rolls of a pair of dice. As it happened, that was not a wise strategy, for the nobleman's

---

*From the correspondence between the two, it's easy to imagine that Fermat solved it with relative ease, not long after Pascal first wrote to him.

arithmetic was slightly off. The probability in that case works out to 49.14 percent, just below evens. Had he instead bet on twenty-five throws, he would have done better since the probability of getting a double 6 in twenty-five throws is 50.55 percent. The moral here is that probability theory can be made to work in the gambler's favor, but you have to do the math right.

The difficulty people had with such calculations was knowing what exactly the numerical odds can and cannot tell you about the future. Many thoughtful gamblers had familiarized themselves with the odds of the various outcomes and knew that the numbers provided useful information about what would happen *on the next throw* (or toss, or whatever). To that extent, they knew that the numerical probabilities provided a form of prediction, but one that was limited to the next round of the game based on their observations that games exhibit regular patterns of outcomes. It was a prediction based, in other words, on their knowledge that, apart from the actual outcome, *one round of the game is exactly the same as any other.*

The problem of the points was significantly different—at least it seemed so in the seventeenth century. The question was not how the next throw might turn out but, rather, if the game had not been abandoned, *how would the players have proceeded?* To us, this seems only marginally different,* but

---

*One way we can look at this today is to imagine a great many continuations of the game and count the relative frequencies of the various outcomes that arise. But such an approach was simply not conceived of before Fermat showed how to do it.

that just reflects how great a shift in perception the Pascal-Fermat correspondence brought about. Reading Pascal's August 24 letter, and seeing his difficulty coming to terms with Fermat's argument, makes it clear that the step the two took in 1654 was a major one.

Yet Pascal and Fermat almost certainly never saw their solution as heralding a major revolution in how humans would view the world. For the two mathematicians, the problem of the points was exactly what it was designed to be: a question about games of chance, a mathematical puzzle relevant to the gambler but not useful in the everyday world. After all, games are not life. A game of chance has a definite number of possible outcomes, each equally likely: two possible outcomes when a coin is tossed, six from the roll of a die, thirty-eight when the casino wheel is spun, and so forth. That permits a precise mathematical analysis of the probabilities.

Using the methods for handling combinations that Pascal described in his book *Traité du triangle arithmétique*, it's a straightforward matter to calculate the total number of possible poker hands, namely, 2,598,960, and from that to compute the probabilities of the various hands: a full house (any three cards of one denomination and any two cards of another) 0.1441 percent, a flush (any five cards of the same suit) 0.1967 percent, and so on. The real world is far less clear-cut. Events happen seemingly at random, and it is only with hindsight, and then only occasionally, that we can look back and see the various alternatives—the other ways things might have turned out. On the face of it, the mathematics Pascal and Fermat developed for games of chance would not work in the messiness and unpredictability of the everyday world. Yet in the very year that Pascal died, a book appeared that would completely change that perception.

THE MAN WHO COUNTED DEAD PEOPLE

John Graunt was born in 1620, the son of a London haberdasher. After serving an apprenticeship in his father's business, he became an employee in the firm and eventually took it over. He grew to be a respected and influential citizen and businessman; held a number of important offices in the Drapers' Company (a London-based international trading company that began by trading woolen cloth in the thirteenth century), in the ward offices, and on the city council; and had friends in cultural and scientific circles. He rose to the rank of major in the London militia. In 1641, he married Mary Scott, and the couple had one son and three daughters. Late in life, he supplemented his fairly basic education with the study of Latin and French. He was a good friend of Samuel Pepys, the famous diarist, and John Aubrey, who included a biography of Graunt in his book *Brief Lives*. It was altogether a highly respectable life and unremarkable, save for one thing. What marked John Graunt as unusual and earned him a revered place in the history books is an eighty-five-page pamphlet published in 1662 and titled *Natural and Political Observations Made Upon the Bills of Mortality.*

In it, Graunt organized and analyzed the mortality rolls of London at that time, in an attempt to help Charles II and other officials to create a system to warn of the onset and spread of bubonic plague in the city. This work marked the beginning of modern statistics. The originality of Graunt's approach was immediately clear to everyone who read his pamphlet, which was not a mere collection of figures. Graunt showed how to reason with data and how to make inferences about a population that led to novel conclusions. Over the ensuing years, it came out in five editions, the second published a few months after the first, and the last in 1676, two years after Graunt died.

The techniques Graunt developed to produce his pamphlet made it possible to move Pascal and Fermat's probability theory out of the gambling room and into the everyday world, providing humankind with a way to control its future fate to a degree completely unimaginable just a few years earlier.

Gaunt was so confident that his pamphlet was important that a month after its appearance, he submitted it in consideration for admission into the Royal Society, the newly formed, august, and highly elite scientific body that would become the model for national learned societies the world over. To buttress his application, Graunt included a dedication to Sir Robert Moray, the president of the Royal Society, and to the other members. In the preface, he referred to a work by the great scientific pioneer Francis Bacon and suggested that his own book should be viewed as a work of natural history, since it concerns "the Air, Countries, Seasons,

Fruitfulness, Health, diseases, Longevity, and the proportion between the Sex and Ages of Mankind." As such, it would, of course, fall squarely within the society's interest.

Still, Graunt seems to have harbored some concern that his background as a mere tradesperson might cause the members to reject his candidacy. (On the pamphlet's title page, he described himself as a "Citizen of London.") So he arranged for support from an unlikely source: no less than King Charles II. It is possible that the king's support was not secured by Graunt directly but through the intervention of his close friend, the surgeon Sir William Petty. In any event, the king obliged and Graunt was duly admitted to the society. Bishop Sprat, the society historian, subsequently wrote of Graunt's admission that

> in his election it was so far from being a prejudice that he was a shopkeeper of London, that His Majesty gave this particular charge to His Society, that if they found any more such tradesmen, they should be sure to admit them all, without any more ado.*

Despite the widespread recognition awarded to Graunt's achievements, he was never able to recover financially after he lost his home and business in the Great Fire of London in 1666, and when he died of jaundice in April 1674, he was a

---

*Bishop Sprat, quoted in *Dictionary of Scientific Biography* (New York: Scribner, 1970–1990).

poor man. After his death, the Drapers' Company provided his widow a pension of four pounds a year "on account of her low condition." It was those who came after Graunt who used his mathematical ideas to become rich.

The focus of Major Graunt's booklet was the bills of mortality that had been collected by the London parishes since 1604. The Company of Parish Clerks published the weekly bills along with an annual bill that summarized the entire year. From 1625, in response to growing public interest, they were printed and made available by subscription of four shillings a year. The impetus for the public's interest was stark and simple: they wanted to know if there had been any deaths in their neighborhood due to the plague. Today, we know that plague is an infectious disease transmitted by fleas carried primarily by rats, gerbils, and squirrels. In Graunt's time, it was believed to be a result of breathing corrupt air. There were only two known methods of defense: quarantine of those unlucky enough to be infected (or believed to be infected) and, for the wealthy, leaving the city for the fresh air of the country until the outbreak passed.

The idea for Graunt's study may have originated with William Petty. In his biography of Graunt in *Brief Lives,* Aubrey wrote:

> He wrote *Observations on the Bills of Mortality* very ingeniosely (but I beleeve, and partly know, that he had his hint from his intimate and familiar friend Sir William Petty), to which he made some *Additions,* since printed.

Regardless of whose idea the study was, no one before Graunt had tried to analyze the masses of data that had been collected. He organized and tabulated the data to show underlying patterns reflecting many aspects of life in London. Much of his effort focused on comparing different entries to adjust for missing figures and inconsistencies. Comparing the causes of deaths from different years, for instance, he was led to conclude that there was significant underreporting and misclassification of cases of plague.

It is not difficult to imagine how this came about. Graunt described how the deaths were classified according to cause by two "Searchers, who are ancient Matrons, sworn to their Office." Since a plague death in a household caused the property to be quarantined, it seemed likely that the ancient matrons, despite being sworn to their office, were not above accepting bribes to make a false entry. "A fourth part more die of the plague than are set down," Graunt wrote.

"French pox" (syphilis) was another cause of death that Graunt concluded was often deliberately misclassified as consumption, ulcers, or sores. In this case, the family of the deceased would have bribed the searchers to avoid the stigma associated with venereal disease.

Graunt often displayed considerable ingenuity in teasing important facts from lists of numbers. When rickets first began to appear as a cause of death, a natural question was whether this was a new disease or simply a new classification for an ailment that had long been in existence. By comparing the *rate* of increase of burials recorded as due to rickets with the rate of burials overall, Graunt concluded that rickets was in fact a new disease with an increasing mortality.

At the time Graunt made his study, it was widely believed that the population of London was around 1 million. Graunt made the first objective estimate, based on three sets of data—births, burials, and numbers of houses—and came up with 460,000, a figure that he revised down to 403,000 in the third edition of his pamphlet, where he used more sophisticated reasoning. He went on to deduce that the population of England and Wales was over 6 million. He arrived at this estimate by noting that London contributed one-fifteenth of the total tax revenue, and hence the entire population must be fourteen times that of the capital, giving the total 14 460,000 = 6,440,000. (Demographers today think the true figure is more likely to have been around 5 million, but acknowledge that Graunt's estimate is remarkable, given the relatively little data on which he based it.)

Graunt also determined that the average family size in London was eight, which seems a large number today but made sense at a time when so many children never lived to be adults. He showed, too, that boys and girls were born in roughly equal numbers—the first time anyone had determined this fact objectively.

Finding that yearly deaths regularly exceeded births, yet the population of London did not decline, Graunt concluded that there was a steady migration from the country into the city. He thus became the first person to document a trend that continues to this day.

A figure of particular interest to the government was the number of men in London of fighting age (defined to be between sixteen and fifty-six). To determine this number, Graunt calculated age-related mortality rates. This marked the introduction of what rapidly became a hugely important concept: the life-expectancy table, the basis for the life insurance industry. Here is how he described what he did:

Having premised these general Advertisements, our first Observation upon the Casualties shall be, that in twenty Years there dying of all diseases and Casualties, 229,250, that 71,124 dyed of the Thrush, Convulsion, Rockets, Teeths, and Worms; and as Abortives, Chrysomes, Infants, Livergrown, and Overlaid; that is to say, that about 1/3 of the whole died of those diseases, which we guess did all light upon Children under four or five Years old.

There died also of the Small-Pox, Swine-Pox, and Measles, and of Worms without convulsions, 12,210, of which number we suppose likewise, that about 1/2 might be Children under six Years old. Now, if we consider that 16 of the said 229 thousand died of that extraordinary and grand Casualty the Plague, we shall finde that about thirty six per centum of all quick conceptions died before six years old.[*]

---

[*]John Graunt, *Natural and Political Observations Made Upon the Bills of Mortality* (1662), 15.

He thus estimated the mortality rate for children aged six years and under to be the number of deaths due to childhood diseases, abortions, and stillbirths, that is 71,124, plus 6,105 (half of the 12,210 due to other diseases except for the plague), divided by the difference between the total deaths (229,250) and those due to the plague (16,000):

$$(71{,}124 + 6{,}105) / (229{,}250 - 16{,}000) = 0.36$$

His reference to "all quick conceptions" was to note that he had included the 8,559 abortions and stillborn among the deaths; if they are excluded, the mortality rate works out at 0.32.

To today's statistician, Graunt's life-expectancy table leaves much to be desired in both accuracy and method. But it was the first attempt in human history at generating such data. Within a short time of the publication of his *Observations,* life-expectancy tables were used widely in medical statistics, demography, and actuarial science, as well as providing the foundation for a rapidly expanding (and still thriving) business of selling life annuities.

Graunt's pamphlet had dramatic effects beyond London. Other European cities, including Paris in 1667, soon introduced bills of mortality. Within a few years, the statistical methods Graunt pioneered had been adopted throughout Europe, leading to the establishment of a number of government statistics offices.

Incidentally, the word *statistics* used to describe the kind of analysis Graunt pioneered comes from the Italian *stato,* meaning "state." A *statista* is someone who deals with affairs of state. But the word *statistics* was not used in England until around 1800. The term the English used before then was *political arithmetic,* so named after William Petty's book *Political Arithmetick,* which he wrote in 1676 but did not publish until 1690.*

After Graunt, all that remained for humanity to gain some control over its future was for the mathematics of calculating probabilities, set in motion by Fermat and Pascal, to be developed sufficiently far that it could be applied productively not just at the gaming tables but also to statistical data about events in the world. That development had already commenced before Graunt ever looked at his first bill of mortality.

A DUTCH TREAT WITH GREAT EXPECTATIONS

In 1657, just three years after the Fermat-Pascal correspondence, there appeared the first account of what is recognizably modern probability theory. The author was the Dutchman Christiaan Huygens, generally regarded as the leading scientist of his day and one of the first members of

---

*Petty's book was influential in stimulating the growth of statistics, but he himself contributed nothing of significance scientifically, so his role in this account is merely that of a messenger.

the Royal Society. His sixteen-page paper *"De ratiociniis in ludo aleae" (On Reckoning at Games of Chance),* a translation into Latin of a paper he had written a year earlier in Dutch, became the standard text in probability theory for the next fifty years.* In it, Huygens established the basic rules for computing probabilities, basing his development on axioms. He was quick to acknowledge that his work built upon the breakthrough made by Pascal and Fermat, among others, writing in his preface that "for some time some of the best mathematicians of France have occupied themselves with this kind of calculus so that no one should attribute to me the honor of its first invention." Huygens went well beyond his two French predecessors in recognizing the potential to apply the methods of probability theory—the term itself had still not come into use—outside the gaming rooms:

> I would like to believe, that if someone studies these things a little more closely, then he will almost certainly come to the conclusion that it is not just a game which has been treated here, but that the principles and the foundations are laid of a very nice and very deep speculation.**

A particularly important concept that Huygens discussed is *expectation* (or *expected gain*). Although he declared that

---

*Huygens's classic pamphlet is reprinted in C. Huygens, *Oeuvres Complètes, Société Hollandaise des Sciences* (La Haye: Nijhoff, 1888–1950).

**Ibid.

none of the work was truly original with him and that he simply built upon the previous work of French mathematicians, in the case of expectation, he is being unduly modest. This concept was at most implicit in earlier work—for instance in Pascal and Fermat's problem of the unfinished game. By recognizing its significance and making it explicit, Huygens took probability theory a huge step forward.

Expected gain is generally regarded as the correct objective measure (in most cases) of the value of a particular wager to the person who makes it. To compute it, you multiply the probability of each outcome by the amount that will be won (or lost, which you count as a negative gain) and add all the results together. For example, casinos offer even odds for betting on red or black at roulette. Suppose you bet $100 on red. The roulette wheel has 36 slots, numbered from 1 to 36, half of them colored red, half black, and two zeros, colored green. The probability of red coming up is therefore 18/38, that is, 9/19. So your expectation (to the nearest cent) is:

$$(9/19 \times \$100) + (10/19 \times -\$100) = -\$100/19 = -\$5.26$$

This means that if you play repeatedly, betting $100 on red each time, then on average, you will lose $5.26 on each game. To put it another way, you can expect your losses to average $5.26 a game.

In Pascal's philosophical work *Pensées*, which he completed in 1658, the great Frenchman used the notion of expectation to argue that one should lead a pious life. We encountered this briefly in Chapter 4. Here is the full argument.

Either God exists or God does not exist. Let p be the probability that God exists; thus $1 - p$ is the probability that there is no God. If you lead a pious life and God exists, you will be rewarded with eternal life in Heaven, which is an infinite reward. If you lead a worldly life and God exists, you will receive a lesser reward, X, that is possibly negative (i.e., a loss). (Some religions warn that X will be an infinite loss—eternal damnation—but Pascal's argument works fine using just the carrot, without the need for a stick.) If God does not exist, then there is no afterlife, so all gains are worldly. Perhaps the (worldly) gain you get from leading a worldly life, Y, exceeds that of leading a pious one full of self-sacrifice, which I'll call Z. (You may even feel that Z is negative.) Your expectation if you lead a worldly life is thus

$$p \times X + (1 - p) \times Y$$

and from leading a pious life, it is

$$p \times \infty + (1 - p) \times Z$$

Since the former is finite and the latter infinite, Pascal concluded, the rational thing to do is lead a pious life. A worldly life is an advantage only if p is zero, but as Pascal said, no one is in a position to know that. For any nonzero probability p of God's existence, no matter how small, piety is the better bet.

Pascal's Wager, as it is called, is cute, and despite the complete absence of any real mathematics, it provides a good example of the basic idea behind modern decision theory. But despite his own deep religious convictions, Pascal did not use his ingenious little scenario to proselytize (as some proselytizers have claimed). Rather, his main conclusion was that "we are compelled to gamble" that God exists.

Indeed we are.

To return to our main story: Sir Robert Moray, the president of the newly formed Royal Society, sent Christiaan Huygens a copy of Graunt's pamphlet when it first came out, but the Dutchman was occupied with many other matters and apparently did not give it more than a cursory look. After completing *"De ratiociniis in ludo aleae"* in 1656, he seems to have written nothing further on probability until his younger brother Lodewijk (Ludwig) approached him in 1669 on the use of Graunt's life-expectancy table for calculating values of life annuities. As things turned out, the two brothers never got to annuities—others took up that challenge—instead becoming engrossed in the prior problem of calculating life expectancy. "The question is to what age a newly conceived child will naturally live," Lodewijk declared in his first letter to his brother.* The younger brother based

---

*Ibid.

his initial investigation on a more extensive life-expectancy table he had drawn up by expanding Graunt's table. He informed Christiaan that by his calculation, the older brother, who was then 40, had 16½ years to live. From the moment Christiaan became involved, the investigation became decidedly mathematical (and one-sided, as Lodewijk was no match for his brother in mathematics).

Since the only framework available for discussing probabilities was games of chance, Christiaan conceived a life table as a lottery having 100 tickets of different values corresponding to the table's entries. He then proceeded to calculate life expectancies using the rule for computing expectation that he had given in his earlier paper. For example, he stated that the number of chances that a person aged sixteen will die before age thirty-six equals 24, and that the number of chances that he will die after age thirty-six equals 16. Thus, in a fair game, you should bet 16 to 24, that is, 2 to 3, that the sixteen-year-old will die before age thirty-six.

If this mathematical exercise strikes you as cold-blooded, you should know that a more sophisticated version of essentially the same calculation is carried out on *your* life expectancy every time you take out a life insurance policy. You probably see life insurance as a way to avoid having to worry about what will happen to your loved ones if you die. The insurance companies see it as a bet on how long you will live. As a business, they have to bet in such a way that their expectation is sufficiently positive for them to make an adequate profit.

The elder Huygens summed up the work in a letter to his younger brother:

> There are thus two different concepts: the expectation or the value of the future age of a person, and the age at which he has an equal chance to survive or not. The first is for the calculation of life annuities, and the other for wagering.[*]

From the standpoint of the ongoing scientific advance that was leading society into today's era of risk management, the crucial factor of the Huygens brothers' collaboration was that it added a layer of abstract mathematics to Graunt's analysis of mortality data.

---

[*]Ibid.

CHAPTER 8

Into the
Everyday World

*. . . I shall not allow myself to disclose to you my reasons
for looking forward to your opinions. I believe you have
recognized from this that the theory of combinations is
good for the case of two players by accident, as it is also
sometimes good in the case of three gamblers, as when one
lacks one point, another one, and the other two, because,
in this case, the number of points in which the game is finished is not enough to allow two to win, but it is not a general method and it is good only in the case where it is
necessary to play exactly a certain number of times.*

In 1654, Pascal had struggled hard to understand why
Fermat counted endings of the unfinished game that would
never have arisen in practice ("it is not a general method
and it is good only in the case where it is necessary to play
exactly a certain number of times"). Just fifteen years later,
in 1669, Christiaan Huygens was using axiom-based abstract
mathematics on top of statistically processed data tables to

determine the probability that a sixteen-year-old young man would die before he reached thirty-six.

But things were about to move even faster. By 1713, the idea that a person could plan his or her life in a rational manner that minimized risk and maximized opportunities, and all with scientific precision, was fully established. In his book *Ars conjectandi (The Art of Conjecture),* published that year, the Swiss mathematician Jakob Bernoulli wrote:

> To conjecture about something is to measure its probability. The Art of Conjecturing or the Stochastic Art is therefore defined as the art of measuring as exactly as possible the probabilities of things so that in our judgments and actions we can always choose or follow that which seems to be better, more satisfactory, safer and more considered.

THE AMAZING BERNOULLIS

"The Amazing Bernoullis" sounds like a family of circus acrobats. It was not, however, on the high trapeze that the Bernoullis performed their dazzling feats, but in mathematics. No fewer than eight members of the Bernoulli family distinguished themselves in the field, and three of them made major contributions to the development of probability theory.

The father of the family was Nikolaus Bernoulli, a wealthy merchant who lived in Basel, Switzerland, from 1623 to 1708. Two of his four sons, Jakob (1654–1705) and

Johann (1667–1748), went on to be first-rate mathematicians.* The other two were Nikolaus (1662–1716) and Hieronymus (1669–1760).

One of Jakob's most significant contributions to mathematics was his "law of large numbers," a major result in the theory of probability. The law of large numbers gives the precise mathematical result that corresponds to the well-known fact that the relative frequency of an event will more accurately predict the likelihood of its occurrence the more trials you observe. (This is why pollsters base their predictions on the opinions of several thousand people rather than a few hundred.)

The other members of the family who contributed to the new mathematics of chance were Johann, Johann's son Daniel (1700–1782), and Jakob's nephew Nikolaus (1687–1759). Johann, Jakob, Daniel, and Nikolaus the nephew were all interested in different aspects of the fundamental question "How can probability theory be taken from the gaming table—where exact probabilities can be calculated—and applied to the far messier real world?"

THE ART OF CONJECTURE

One of the problems Jakob worked on was already implicit in Graunt's work on mortality rates. Graunt was well aware

---

*German, French, and English versions of the Bernoullis' names are common in the literature, reflecting the truly international nature of the impact their work had on the world.

that the data he had, while extensive, still represented only a sample of the population of London. Moreover, the figures were for a specific period. Yet this did not prevent the Englishman from making generalizations that went beyond the data itself. How much can we rely on the conclusions obtained from the study of a sample? Specifically, if a probability is computed on the basis of a sample, how reliable can we take that probability to be? Will a larger sample ensure a more reliable result, and if so, how large does the sample have to be? Jakob Bernoulli raised this very question in a letter to his friend Gottfried Leibniz (one of the two inventors of calculus) in 1703.

In his pessimistic reply, Leibniz observed that "nature has established patterns originating in the return of events, but only for the most part." It was that phrase "only for the most part" that seemed to stand in the way of a mathematical analysis. Referring to mortality rates, which Bernoulli had given as a specific example, Leibniz went on to say that "new illnesses flood the human race, so that no matter how many experiments you have done on corpses, you have not thereby imposed a limit on the nature of events so that in future they could not vary."*

Undeterred by Leibniz's reply, Bernoulli continued his investigation, and in the two years that remained to him, he made considerable progress. (After he died, in 1705, his

---

*Gottfried Leibniz to Jakob Bernoulli, quoted in John Maynard Keynes, *A Treatise on Probability* (London: Macmillan, 1921).

nephew Nikolaus Bernoulli began organizing his uncle's results into publishable form, a task so challenging that it took
him eight years before *Ars conjectandi* was published.)

Jakob began by recognizing that the question he had
asked Leibniz really split into two cases, depending on two
notions of probability. First, there was what Jakob called *a
priori* probability—probability computed before the fact.*
Was it possible to compute a precise probability of the outcome of an event before the event occurred? With games of
chance—cards, dice, and so forth—the answer was yes. But
as Leibniz rightly pointed out, probabilities computed in advance for events such as illness or death could be reliable
"only for the most part."

Jakob used the term *a posteriori* probability to refer to
the other kind of probability—probability computed after
the event. Given a sample of a population, if a probability is
computed for that sample, how accurately does that probability represent the entire population? To illustrate, suppose
you were given a large, opaque jar of red and blue marbles.
You know there are 5,000 marbles in all, but you don't know
how many there are of each color. You draw one marble
from the jar at random, and it is red. You put it back, shake
up the jar, and then draw again. This time, you draw a blue

---

*Bernoulli's notions of *a priori* and *a posteriori* probabilities, which I describe
here, are not the same as the *prior* and *posterior* probabilities that arise in
Bayes' theorem, which we'll look at in the next chapter. This simply adds further confusion to an area already mired in confusion, I know, but we are stuck
with the terminology.

marble. You repeat the process—drawing, returning, shaking, drawing—50 times and find that 31 times you draw a red marble and 19 times a blue marble. This leads you to suspect that there are around 3,000 red marbles and 2,000 blue ones, so that the *a posteriori* probability of drawing a red marble at random is 3/5. How confident can you be in this conclusion? Would you be more confident if you had sampled more marbles—say, 100?

Bernoulli proved that by taking a sufficiently large sample, you can increase your confidence in the computed probability to whatever degree you wish. More precisely, by increasing the sample size, you can increase to any level your confidence that the probability computed for the sample is *within any stipulated amount* of the true probability. This is the law of large numbers.

In the case of the 5,000 marbles in the jar, where there are exactly 3,000 red ones and exactly 2,000 blue, Bernoulli calculated how many marbles you would have to sample to be certain to within 1 in 1,000 that the distribution found in the sample is within 2 percent of the true ratio of 3 to 2. The answer he obtained was 25,550 drawings. That's far more than the total number of marbles (5,000). So, in this example, it would be much more efficient simply to count all the marbles! Nevertheless, Bernoulli's theoretical result shows that by taking a sufficiently large sample, probabilities *can* be computed from a sample so that to any degree of certainty (other than absolute certainty), the computed probability is within any desired degree of accuracy of the true probability.

With *a posteriori* probability, there is a second question: How reliable a guide is such a probability for predicting future events? This is not really a mathematical question. The issue is rather how much the past can be taken as indicative of the future. The answer can vary. As Leibniz remarked, new illnesses flood the human race, so that no matter how many experiments you have done on corpses, you have not thereby imposed a limit on the nature of events. More recently, the late Fischer Black, a pioneering financial expert who exchanged his academic position at MIT for the finance market on Wall Street, made a similar remark about basing predictions on mathematical analysis of past events: "Markets look a lot less efficient from the banks of the Hudson than the banks of the Charles."*

Probability Goes to Court

The new science of quantifying the likelihoods of everyday events found its first applications in the law. While working to bring order to his uncle Jakob's *Ars conjectandi* and make it suitable for publication, Nikolaus Bernoulli made significant contributions of his own. In 1709, he published a book titled *De usu artis conjectandi in jure* (*The Use of the Art of Conjecturing in Law*). In Chapter 2, which is prefatory to his later examination of legal issues, Nikolaus discusses the

---

*Fischer Black, quoted in Peter Bernstein, *Against the Gods* (New York: Wiley, 1996), p. 7.

"estimation of the length of human life." In particular, he sets out to justify his Uncle Jakob's statement (made without any proof in a 1686 publication) that the odds of a sixteen-year-old dying before a fifty-six-year-old are 59 to 101. (In modern terms, we would say that the probability that the younger person will die first is 59/(59 + 101) = 59/160, or about 36.9 percent.)

Nikolaus obtained the same answer by applying probability theory (specifically, the methods of Huygens and his Uncle Jakob) to Graunt's life table, which was presumably how his uncle had done it. Following Jakob's lead, he noted that the probability of surviving to a certain age cannot be found by *a priori* reasoning as in a game of chance but may be estimated by observation—say, by observing how many of three hundred men survive for ten years. Nikolaus noted that this is precisely how Graunt constructed his life table: "[It] has been observed from collation of very many catalogues of this sort."

One of the intriguing legal problems Nikolaus considered in his book is how much time must elapse after an individual has "gone absent," with no indication that he or she is alive or dead, before a court may legally declare that person dead. This question arises all the time in court cases involving inheritance or widowhood.

After discussing the legal meaning of the term *absent*, Nikolaus quotes a number of prominent jurists who hold widely differing opinions on how long one should wait, ranging from five years to thirty. He proposed to resolve the matter by applying probability theory to Graunt's life table. His idea was to find "in what time it is twice, thrice, four times etc. as likely that someone be dead than that he is still alive." He solved the problem completely for the case where the probability of being dead is twice that of being alive, producing this table (which I have rounded to the nearest whole numbers):

Age when last seen:	0	6	16	26	36	46	56	66	76
Time to wait (years):	21	24	25	25	23	20	15	10	7

Nikolaus concluded that

if someone should have started his absence at the twentieth or thirtieth years of his life, for example, and has been absent for 25 years, and nothing has been heard from him through that time, the Judge would be able to declare him dead and to grant his goods to his nearest relatives without caution.

A NEW KIND OF PROBABILITY

It is important to recognize that the new kind of probability introduced by Jakob Bernoulli is not the same as the older one. *A priori* probability, which applies at the gaming table, depends on the symmetry of the game—a die is equally likely to land with any one of its six faces up. It is an objective fact associated with the game and does not depend on the observer. The *a posteriori* probability that Bernoulli introduced was something radically different: a measure of an observer's knowledge of the truth of a proposition—a numerical "degree of certainty." If two people compute the *a posteriori* probability of (their knowledge of) the same event, they may come up with different answers. Moreover, those answers can change if they acquire additional information about the event. Many people today find probability confusing because they fail to recognize this distinction.

For the most part, it is *a posteriori* probability that we use in predicting the future in the everyday world. Thus, Jakob Bernoulli's introduction of this new notion marked a significant step toward risk management.

Notice that, from a purely mathematical perspective, you can regard *a priori* probability as a special case of Bernoulli's notion. For example, if you are being rational, you should assign the probability 1/6 to your "knowledge" that a die about to be thrown will land 6 up. Thus, it is not really true that mathematicians found a way to apply their existing concept of probability to the everyday world. Rather, they developed a new concept that is intrinsically about the everyday world, which in terms of the values it yields can be viewed as an extension of the old concept, and which enabled them to apply to the everyday world the mathematical methods they had developed to quantify the outcomes in games of chance.

It was at this moment that the term *probability theory* came into use to describe the mathematics that had been pursued by Pacioli, Cardano, Pascal, Fermat, and Huygens.

Focusing on problems of games of chance, those pioneers had all referred to the concept they were studying as *chance*. In their day, *probability* was a nontechnical word that referred to uncertain events in the future, with certain eventualities being viewed as "probable" or "improbable," but with no expectation that such judgments could be rendered mathematically precise. Following Bernoulli's use, the word *probability* shifted from reflecting uncertainty about the future to denoting the mathematical method for making specific numerical predictions.

CHAPTER 9

The Chance of Your Life

. . . Consequently, as you did not have my method when you sent me the division among several gamblers, but [since you had] only that of combinations, I fear that we hold different views on the subject.

I beg you to inform me how you would proceed in your research on this problem. I shall receive your reply with respect and joy, even if your opinions should be contrary to mine.

I am etc.

And so Pascal ends his long letter.* Although the subsequent development of probability theory was rapid, it might have been even faster if, shortly after Pascal and Fermat had their exchange of letters, Isaac Newton and Gottfried Leibniz had not invented calculus, the mathematics of continuous

---

*We saw Fermat's August 29 reply in Chapter 6.

motion and change. To live at a time when new mathematics that will change the world is being invented is a rare privilege granted to few generations. For the mathematicians in the second half of the seventeenth century, there was not one such revolution going on but two. Although the social impacts of both calculus and probability theory were huge, calculus offered by far the greater challenge to mathematicians, since the mathematics was much more exceptional. As a consequence, most of the focus of the greatest mathematical minds was on the development of calculus. Probability theory took second place. Nevertheless, progress was steady and dramatic, as mathematicians not only continued to develop the mathematics of probability itself but also found ways to apply it to everyday life.

Jakob Bernoulli's introduction of *a posteriori* probability was a major advance. The distinction between that and the older "dice game" probability was the first step toward today's notions of *subjective* and *objective* probability.* Graunt's life tables and Nikolaus Bernoulli's use of probabilities in courts of law provided a numerical measure of the reliability of an individual's knowledge of a particular event, but those probabilities were objective. What classified them as *a posteriori* (rather than *a priori*) was that the probability

---

*The distinctions between the various kinds of probabilities are not at all clear-cut and should not be thought of as clean, scientific categories. Mathematicians avoid the issue of what exactly probability is by developing it in an axiomatic fashion. In mathematics, *probability* is what the axioms say it is, period.

values assigned to various events were determined not by reasoning logically from symmetry conditions, as in a game of dice or cards, but by (the *objective* process of) counting outcomes of trials in the world. *Subjective* probability takes this a step further.* One of the first people to attempt to capture mathematically what people do when they try to make rational decisions about the future was another Bernoulli: Johann's son Daniel.

FEAR OF FLYING

As friends of Leibniz, Jakob and Johann Bernoulli were among the first mathematicians to learn about the remarkable new method of calculus. In correspondence with Leibniz, Jakob and Johann carried out much of the early development of the subject, and many of the techniques learned by today's calculus students were developed by the two Bernoullis. (Some historians suggest that the rich subject we today call calculus was created by not two but four mathematicians: Newton, Leibniz, and the two Bernoullis.) Daniel Bernoulli, too, did pioneering work on calculus, but he took a different course from his Uncle Jakob's, concentrating on applying the methods of calculus to flowing liquids and gases. Arguably the most significant of Daniel's many discoveries was what is now known as Bernoulli's equation, which describes one of the sources of lift that

---

*The previous footnote still applies.

keep an aircraft aloft.* Though the airline industry's use of this equation was two hundred years away, other work of his was also relevant to the future world of air travel. Daniel's contributions to our understanding of probability help explain the oft-cited observation that, although modern air travel is the safest way to travel, many people nevertheless feel extremely nervous boarding a plane. Some people fear flying so much that they never get on a plane at all.

Such people may well know that the probability of being involved in a major accident is far less than when they travel by car. What scares them is the often horrific nature of an airline crash and the importance they attach to such an event, however unlikely.

The fear of lightning is similar. The tiny mathematical probability of being struck by a lightning bolt is far outweighed by the significance many people attach to it.

It was this essentially human aspect of probability calculations that interested Daniel Bernoulli. Was it possible, he asked, to make specific observations about the way people actually assessed risk? In 1738, he published a seminal paper on the issue in the *Papers of the Imperial Academy of Sciences in Saint Petersburg.*** In that paper, he introduced the

---

*There are three main sources of lift for a modern aircraft: the Bernoulli effect, turbulent air flow around the aircraft, and, the most significant of the three, the plain old Newton's Third Law reaction force that results from air being forced downward over the fuselage and wings.

**Reprinted in D. Speiser, ed., *Die Werke von Daniel Bernoulli* (Basel: Birkhäuser, 1982).

key new concept of *utility*, which he based on Huygens's notion of expectation.

By taking account of both the probabilities and the payoffs, expectation was supposed to measure the value of a particular risk or wager, from an individual's perspective. The greater the expectation, the more attractive the risk. That, at least, was the theory. For many examples it seemed to work well enough. But there was a problem, most dramatically illustrated by a tantalizing puzzle, proposed by Daniel's cousin Nikolaus, that came to be known as the Saint Petersburg paradox. Here it is:

Suppose I challenge you to a game of repeated coin tosses. If you throw a head on the first toss, I pay you $2 and the game is over. If you throw a tail on the first throw and a head on the second, I pay you $4 and the game is over. If you throw two tails and then a head, I pay you $8 and the game is over. We continue in this fashion until you throw a head. Each time you throw a tail and the game continues, I double the amount you will win when you eventually throw a head.

Now imagine that a friend comes along and offers to pay you $10 to take your place in the game. Would you accept or decline? What if your friend offered you $50? Or $100? In other words, how much do you judge the game to be worth to you?

This is exactly what expectation is supposed to measure. So what does it work out to be in this case? Well, in principle, the game can go on indefinitely—there are infinitely many possible outcomes: H, T H, T T H, T T T H,

T T T T H, and so forth. The respective probabilities of these outcomes are 1/2, 1/4, 1/8, 1/16, 1/32, and so forth. Thus, the expectation (i.e., your expected reward) is

$$(\tfrac{1}{2}) \times 2 + (\tfrac{1}{4}) \times 4 + (\tfrac{1}{8}) \times 8 + (\tfrac{1}{16}) \times 16 + \ldots$$

This infinite sum can be rewritten as

$$1 + 1 + 1 + 1 + \ldots$$

Since the sum goes on forever, the expectation is therefore infinite.

According to the theory, faced with an infinite expectation, you should not give up your opportunity to play for any amount of money. But most people—even probability theorists—are tempted to take the $10 offer and would almost certainly take $20. Why?

Pondering this question, as well as a number of other problems he saw with expectation, led Daniel Bernoulli to replace the highly mathematical concept of expectation with the far less formal idea of utility. Utility is intended to measure the significance of a particular outcome to a specific individual. As such, it is very much a personal thing. It depends on the value someone puts on a particular event. Your utility and mine might differ.

At first glance, the move to replace the mathematically precise concept of expectation with the decidedly personal idea of utility might appear to render further scientific analysis impossible. But it doesn't. Even for a single individual, it may well be impossible to assign specific numerical values to utility. Yet Bernoulli was able to make a profound observation about utility. He wrote that the "utility resulting from any small increase in wealth will be inversely proportionate to the quantity of goods previously possessed." Bernoulli's utility law explains why, even for a moderately wealthy person, the pain of losing half of one's fortune is much greater than the pleasure or benefit of doubling it. Thus, few of us are prepared to gamble half our wealth for the chance of doubling it. Only when we are truly able to declare "What have I got to lose?" are most of us prepared to take a big gamble.

For example, suppose you and I each have a net worth of $10,000. I offer you a single toss of a coin. Heads, I give you $5,000; tails, you give me $5,000. The winner comes out with $15,000, the loser with $5,000. Since the payoffs are equal and the probability of each of us winning is 1/2, we each have an expectation of zero. In other words, according to expectation theory, it makes no difference to either of us whether we play or not. But most of us would not play, because we would view the risk of losing unacceptable. The 0.5 probability of losing $5,000 (half our wealth) far outweighs the 0.5 probability of winning $5,000 (an increase of 50 percent).

The utility law likewise resolves the Saint Petersburg paradox. The longer the game goes on, the greater the amount you will win when a head is finally tossed. (If the game goes on for six tosses, you are bound to win over $100. Nine tosses, and your winnings will exceed $1,000. If it lasts for fifty tosses, you will win over $1 million billion.) According to Bernoulli's law, once you reach the stage where your minimum winning represents a measurable gain *in your terms*, the benefit to be gained by playing longer starts to decrease. That determines the amount for which you would be prepared to sell your place in the game.

So utility superseded expectation. A similar fate eventually befell Bernoulli's utility concept when mathematicians and economists looked more closely at human behavior. Still, it was Daniel Bernoulli who first insisted that if you wanted to apply the mathematics of probability theory to real-world problems, you had to take account of the human factor.

What made risk management a mathematical science rather than a branch of psychology—although it undoubtedly has an element of art to its application—is that the human factor is itself amenable to mathematical analysis. Leading this development was none other than a Bernoulli, though a Frenchman living in England would make the real breakthrough.

RINGING IN THE BELL CURVE

Jakob Bernoulli had shown how to determine the number of observations needed to ensure that the probability determined by the sample was within a specified amount of the true probability. This was of theoretical interest, but of little practical use. For one thing, it required that you know the true probability in advance. Second, as Bernoulli had himself shown in the case of the marbles in the jar, the number of observations required to achieve a reasonably accurate answer could be quite large. Potentially much more useful was the opposite problem: Given a specified number of observations, can you calculate the probability that they will fall within a specified bound of the true value?

The first person to try to investigate this question was Jakob's nephew Nikolaus, who worked on it while completing his late uncle's work and preparing it for publication. To illustrate the problem, Nikolaus gave an example concerning births. Assuming a ratio of 18 male births to every 17 female births, for a total of 14,000 births the expected number of male births would be 7,200. He calculated that the odds were over 43 to 1 that the actual number of male births in the population would lie between 7,200 − 163 and 7,200 + 163—that is, between 7,037 and 7,363. Nikolaus did not completely answer the question, but he made sufficient progress to publish his findings in 1713, the same year his deceased uncle's book finally appeared.

Some years later, Nikolaus's ideas were taken up by Abraham de Moivre (1667–1754), a French mathematician who, as a Protestant, had fled to England in 1688 to escape persecution by the Catholics. Unable to secure a proper academic position in his adopted country, de Moivre made a living by

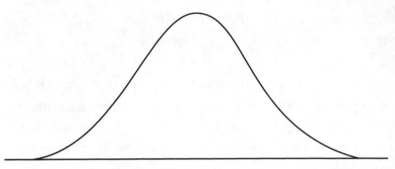

FIGURE 1. The Normal Distribution

tutoring in mathematics and—a sure sign of the growth of probability theory for risk management—consulting for insurance brokers.

Where Nikolaus Bernoulli had made only partial progress on his problem, de Moivre obtained a complete solution, which he published in 1733 in the second edition of his book *The Doctrine of Chances*. Using methods of calculus as well as probability theory, de Moivre showed how a collection of random observations would distribute themselves around their average value. These days, this distribution is known as the *normal distribution*. When it is represented graphically, plotting observations along the horizontal axis and the frequency (or probability) of each observation along the vertical axis, the resulting curve is shaped like a bell. For this reason it is often called the *bell curve* (see Figure 1).

The bell curve shows that most observations are clustered in the center, around the mean value. Going out from the middle, the curve slopes down symmetrically, with an equal number of observations on both sides of the mean. At first it falls slowly, then much more rapidly, finally becoming very flat toward the ends. In general, observations far from the mean are less frequent than observations close to the mean.

De Moivre discovered the bell curve by examining the behavior of random observations. Its elegant, symmetrical shape showed that beneath randomness lay an elegant geometry. From a mathematical point of view, that alone was a significant result. But that was not the end of the story. Eighty years later, the great German mathematician Karl Friedrich Gauss noticed that whenever he plotted a large number of measurements of something, the resulting curve always looked remarkably like de Moivre's bell curve. Different measurements of a particular distance on the earth's surface, for example, or of an astronomical distance, would form a bell-shaped cluster around a central value—the average of the individual measurements. The inevitable errors in taking measurements gave rise, it seemed, to a normal distribution of values.

By considering the bell curve not as a geometric feature of randomness but as an inevitable consequence of errors in measurements, Gauss realized that he could use the normal distribution to provide a framework for assessing the value of data. In particular, using the bell curve, it would be possible to assign (approximate) numerical probabilities to events, much as they could be assigned to outcomes of a roll of dice or a spin of the roulette wheel. The closer a particular datum was to the mean, as measured on the bell curve, the greater the probability that it was correct. The techniques of probability theory developed by Pascal, Fermat, Huygens, the Bernoullis, and others could then be transferred from the gambling table to other areas of life. As a result of Gauss's contribution, the bell curve is sometimes called the *Gaussian distribution*.

The key technical concept Gauss needed to use the bell curve to assess the reliability of data was a measure of the dispersion around the mean that de Moivre himself had introduced. Known nowadays as the *standard deviation*, de Moivre's measure allows us to judge whether a set of observations is sufficiently representative of the entire population. Although based on some solid mathematics, standard deviation is a somewhat arbitrary notion that gives a numerical answer to the question "How much do the individual data points differ from the mean?" For a normal distribution, approximately 68 percent of the observations fall within one standard deviation of the mean and 95 percent fall within two standard deviations of the mean. For this reason, when newspapers and magazines publish results of surveys, they generally include a statement of the deviation. If they don't, you should view the results with suspicion. Remember, a man whose head is in an oven and whose feet are encased in ice could be said to feel just fine on average. But what a deviation!

With statistics playing such a major role in twenty-first-century life, the bell curve is an icon of our age. It allows us to assign numerical probabilities to events and thereby apply the methods of probability theory. In general, whenever there is a large population that produces data, and the data produced by each member of the population is independent of that produced by the others, then the bell curve will arise and Gauss's method provides a measure of the reliability of that data.

Following Gauss's lead, business leaders now use the bell curve to plan expansions, officials use it to determine public policy, educators use it to grade student examinations, pollsters use it to analyze their data and make their predictions, medical researchers use it to test the effectiveness of different treatments, economists use it to analyze economic performance, biologists use it to study plant growth, psychologists use it to analyze the behavior of their fellow humans, and insurance companies use it to set their premiums.

I'll look at this last application as an example.

The Price on Your Head

Few people would classify themselves as regular gamblers— in the sense, say, of frequenting casinos or participating in weekly poker games—yet a great many buy tickets in state lotteries. In so doing, they are gamblers in the technical sense of betting money in the hope of winning something, but they generally do not see themselves as gambling. They pay a small fixed fee, and their subsequent involvement is at most that of checking the list of winning numbers. Moreover, the game they are playing (in fact, it's hardly a game at all) has fixed odds and there is limited opportunity to secure an advantage. Most people who buy tickets are not aware of or even interested in the odds against winning—perhaps because the probability of winning for any single individual is essentially zero. Thus, the typical buyer of a lottery ticket is unaffected by probability theory; his or her odds are effectively zero, and winning is a matter of pure luck.

The same is not true for the four areas where a great many people "gamble" every day: life, health, property, and savings. Apart from investing in the stock market, these are all a matter of taking on risks rather than gambling per se. But most of us—or the companies we deal with, such as the bank that lends us money—try to mitigate undesired consequences of these risks by taking out some form insurance or annuity. That is very definitely gambling, and the companies that assume our risk have to do the math right, to be able to charge us a fee that enables them to stay in business while being acceptably fair to their customers.

The oldest example of this form of gambling is the life annuity, which dates back at least to the Roman Empire. The person who takes out a typical annuity is not really gambling: you simply pay a fixed amount by a specified date in order to receive regular payments for the remainder of your life. But to the authorities or, these days, the companies that sell you the annuity, it is very much a gamble: they are betting on your life. If you die before they have paid out to you the full value of your purchase price, they have made a profit on you; if you live a long time, they take a loss. Clearly, their task is to set the purchase price so that, when averaged over all their customers, they make an acceptable profit while avoiding charges of profiteering on people's misfortunes. There are generally laws that stipulate a minimum number of payments that must be made out to the heirs of an individual who dies before a fixed amount of time has elapsed.

Such laws existed in ancient Rome, where around 220 A.D., the jurist D. Ulpian devised the following table for converting a life annuity to one with a fixed number of payments.*

AGE OF ANNUITANT	0–19	20–24	25 29	30–34	35–39	40 ... 49	50–54	60–
DURATION OF ANNUITY (YEARS)	30	28	25	22	20	19 ... 10	9	5

It is not known whether this table was based on observational data or was simply a legal stipulation, but the figure of thirty years' duration for a person who dies before the age of twenty, in an era when the life expectancy was much lower, suggests the latter. Nikolaus Bernoulli discussed Ulpian's table in his book *De usu artis conjectandi in jure.* He carried out his own computation based on Graunt's life table and came up with a figure of twenty-one years, instead of thirty,

---

Corpus Iuris Civilis, Digesta, 35: 2, 68.

for conversion when the annuitant is less than twenty years old at the time of death. Nikolaus examined another annuity system, offered by the magistrate of Amsterdam in 1673, and declared that the prices charged were close to those he himself had calculated.

In the Middle Ages, it was common practice for states and towns to sell annuities to raise funds. In general, the price was not dependent on the age of the annuitant, and as a result, buyers generally nominated healthy children. In his book, Bernoulli quoted an announcement made in London in 1704:

> Yesterday the Queen issued an order for the sale of life annuities; in two hours £10,000 were subscribed and to date £100,000. These are the conditions for the annuities; if someone wants 10 pounds of yearly income with a possible extension to two succeeding persons, then he pays 90 pounds for one life, 100 pounds for two lives, 120 pounds for three lives following one another; and if one wishes to have 14 pounds yearly for 99 years, one pays 210 pounds for it.

A detailed study of annuities was made by a remarkable Dutchman by the name of Jan de Witt (1625–1672), who, after a brilliant early career as a mathematician, went into politics and became, at age twenty-eight, the prime minister of Holland. In need of funds to fight the English and French, he proposed the sale of annuities and wrote a paper that described how to price and structure them, based on an original and sophisticated mathematical analysis. The paper was published in 1671, one year before France's invasion of Holland, an event that led to his resignation from office and subsequent murder by a mob.

Soon after de Witt's work, the famous English scientist Edmund Halley (1656–1742) made further progress. Halley was an immensely talented and versatile scientist who did leading work in astronomy, geophysics, physics, and mathematics. He is best known today for predicting the return of a now eponymous comet, but he also wrote a paper on insurance mathematics, which he published in the Royal Society's scholarly publication *Philosophical Transactions,* of which he was the editor. Like everything else he did, it was first-rate.

Halley's study was a mathematically more sophisticated version of what John Graunt had done thirty years earlier. Halley's sources of data were the records of all births and deaths, listed according to gender and age, that had been kept in Breslau, in Silesia (in modern-day Poland), since the end of the sixteenth century. After Graunt's work, the Royal Society had been eager to locate another source of data to construct a life table, particularly one that was more complete and hence more reliable. When members heard of the Breslau records, which were indeed far better than those available to Graunt, they asked Halley to analyze them. This he duly did, presenting his results to the society in a 1693 paper with the pithy title *An Estimate of the Degrees of the*

Mortality of Mankind, drawn from curious Tables of the Births and Funerals at the City of Breslaw; with an Attempt to ascertain the Price of Annuities upon Lives.

Unlike Graunt, Halley was a mathematician of great power, and his analysis was far more sophisticated than the one the major had carried out. The tables he produced provided the basis for offering annuities and life insurance policies for many years. Among many findings, Halley calculated the annual death rate and infant mortality and found that both were close to those obtained by Graunt, thereby supporting the belief that Graunt's figures were reliable.

With insurance in particular, ordinary people were able to take advantage of the ability that probability theory gives us to foresee the future (or at least quantify what it might bring). An insurance company cannot say with any certainty that you will live long or avoid major injury, but it can determine with considerable certainty what the various likelihoods are for society as a whole, or for the subgroup of "people like you," and that enables it to make a living by betting on your chances—just as a casino makes its profits not on one customer's luck but on the aggregate performance of all its customers.

With Halley's work, probability theory completed its migration from the gaming rooms to the everyday world, where it could be used to make predictions about the future. Initially, those predictions were restricted to things that could be computed from aggregate data, like the expected age at which a person would die. But things were about to take a dramatic turn. The modern era was about to dawn.

THE REVEREND BAYES

In the case of insurance and similar applications of probability theory, prediction of the future is reliable only on average, over a large number of cases of essentially the same type. But today we use probability another way: to measure the likelihood of our being correct when we try to predict *one particular event*. Although it had to await the computer age to come to fruition, the final mathematical* step in the path from the solution of the problem of the points to today's risk-managed society was an ingenious and extremely powerful mathematical formula developed by an obscure eighteenth-century Presbyterian minister in England who studied mathematics as a hobby.

Thomas Bayes was born in London in 1702. His father, Joshua, was one of the first Nonconformist ministers to be ordained in England, which meant that his son was not allowed to study at Oxford or Cambridge. Accordingly, at seventeen, Thomas headed north to Scotland to enroll at the University of Edinburgh, where he studied logic and theology. In due course, he was ordained like his father and, in

---

*This book concentrates on the mathematical developments; there were several other advances and innovations that played roles in acquiring the ability to manage risk.

1733, became a minister in the Presbyterian Chapel in Tunbridge Wells, a well-to-do market town some thirty-five miles southeast of London. He served there until he retired in 1752, after which he remained in the town until his death, in 1761.

Bayes is recognized today as having a brilliant mathematical mind, but he published no original mathematics during his life. (After his death, a private notebook was found containing discussions of probability, trigonometry, geometry, solution of equations, series, differential calculus, electricity, optics, and celestial mechanics.) In 1736, he did publish *An Introduction to the Doctrine of Fluxions, and a Defence of the Mathematicians Against the Objections of the Author of The Analyst,* a predominantly philosophical article about Newton's calculus, rebutting Bishop Berkeley's attack on the logical foundations of Newton's method—but that seems to be all. Despite his lack of publications, his scientific abilities must have been known to others, for in 1742 he was elected a Fellow of the Royal Society.

When Bayes died, he bequeathed his papers to his friend Richard Price, a talented mathematician and also a Fellow of the Royal Society, who would go on to do foundational work in what later became actuarial science. Among the documents Price inherited, one item in particular caught his attention. Its title was *Essay towards solving a problem in the doctrine of chances,* and it outlined a radically new way to approach and compute probabilities. Recognizing its importance, Price submitted it to the Royal Society for publication with the following cover note:

> I now send you an essay which I have found among the papers of our deceased friend Mr Bayes, and which, in my opinion, has great merit. . . . In an introduction which he has writ to this Essay, he says, that his design at first in thinking on the subject of it was, to find out a method by which we might judge concerning the probability that an event has to happen, in given circumstances, upon supposition that we know nothing concerning it but that, under the same circumstances, it has happened a certain number of times, and failed a certain other number of times.*

Other scholars at the Royal Society also felt Bayes' essay had merit, and it was duly published in the *Philosophical Transactions* in 1764.

The importance of Bayes' paper is not merely the formulation of the specific problem Price referred to in his cover letter, but the revolutionary and widely applicable approach to probability that Bayes adopted to solve the problem. The probabilities his method computes are probabilities attached to a person's knowledge of some event—they tell you the degree of confidence in some piece of information. In

---

*Richard Price, in M. G. Kendall and R. L. Plackett, *Studies in the History of Statistics and Probability*, vol. 2 (New York: Macmillan, 1977).

that respect, Bayes was following the work of Jakob Bernoulli, but he took things considerably further. (Unfortunately, he used the confusingly similar terms "prior probability" and "posterior probability," but with different meaning from Bernoulli's *a priori* and *a posteriori,* so it is best to view Bayes' approach purely on its own.)

Bayes' method does not tell you how to calculate the probability of a hypothesis *de novo*. Rather, it is a method for *revising* a probability in light of new information. You start with a figure for the probability of a hypothesis, H; that figure is called the *prior probability* for the hypothesis H. Given some new information E, you then carry out a calculation to obtain a revised probability for H; that new figure is called the *posterior probability*. The revision is carried out by inserting the relevant figures into a mathematical formula known as Bayes' formula (or sometimes Bayes' rule).

The prior probability might be a guess or an estimate. Given sufficient new information, an application of Bayes' revision procedure can lead to a more accurate probability. By applying Bayes' method iteratively (generally using a computer) each time new information is obtained, one may turn even a fairly poor prior probability into a fairly reliable posterior probability. (On the other hand, the method is not immune to the familiar adage about excessive reliance on computer models: garbage in, garbage out.)

Because the method depends on an initial "seed" value for the (first) prior probability, for two hundred years after

Bayes' method became known, it was largely ignored by the statistics and probability theory communities. Starting in the 1970s, however, it has grown in popularity, in large part because the availability of powerful computers has made it possible to run the process iteratively with large amounts of information, often overcoming the inaccuracy of a poor first prior.

Bayes' Formula

For the record, I'll describe Bayes' formula to calculate the probability of a certain hypothesis H, based on evidence E. (If mathematical formulas give you the willies, just skip to the next chapter, though if you do, you'll miss an opportunity to learn how to interpret a seemingly devastating medical prognosis.)

Let $P(H)$ be the probability that the hypothesis H is correct in the absence of any evidence—the *prior probability*. This is the figure you start with. Where you get it from is your business.

Let $P(H|E)$ be the probability that H is correct, given E. This is the revised estimate you want to calculate. A quantity such as $P(H|E)$ is known as a *conditional probability*—the conditional probability of H occurring, given E.

Let $P(E|H)$ be the probability that E would be found if H were correct, and let $P(E|H_{wrong})$ be the probability that E would be found if H were false. Bayes' formula requires these figures.

To compute the new estimate, you also have to calculate $P(H_{wrong})$, the probability that H is false. This is easy: the probability that H is wrong is the same as 1 minus the probability that H is correct: $P(H_{wrong}) = 1 - P(H)$.

Bayes' formula is:

$$P(H/E)= \frac{P(H) \times P(E|H)}{P(H) \times P(E|H) + P(H_{wrong}) \times P(E|H_{wrong})}$$

One of the strengths of Bayes' approach is that it can sometimes guide us when our intuitions are wrong. For example, suppose you undergo a medical test for a relatively rare cancer. The cancer has an incidence of 1 percent among the general population. Extensive trials have shown that the reliability of the test is 79 percent. More precisely, although the test does not fail to detect the cancer when it is present, it gives a positive result in 21 percent of the cases where no cancer is present—what is known as a false positive. When you are tested, the test produces a positive diagnosis. The question is, What is the probability that you have the cancer?

If you are like most people, you will assume that if the test has a reliability rate of nearly 80 percent, and since you tested positive, the likelihood that you do indeed have the cancer is about 80 percent (i.e., the probability is approximately 0.8). Are you right?

No. Given the scenario I just described, the likelihood that you have the cancer is just 4.6 percent (i.e., the probability is 0.046). There is a less-than-5-percent chance that you have the cancer. Still a worrying possibility, of course. But hardly the scary 80 percent you thought at first.

Here is how you arrive at that figure:

$P(H)$ = 0.01 (the cancer has a 1 percent incidence in the population)

$P(E|H)$ = 1 (the test always shows positive if the cancer is present)

$P(H_{wrong})$ = 0.99 (99 percent of the population is cancer-free)

$P(E|H_{wrong})$ = 0.21 (the test gives a false positive in 21 percent of cases)

So by Bayes' formula

$$P(H/E)= \frac{0.01 \times 1}{(0.01 \times 1) + (0.99 \times 0.21)} = \frac{0.01}{0.2179} = 0.0459$$

Bayes' formula comes from a fairly straightforward enumeration and counting argument, much like the one Fermat used to solve the problem of the points. Rather than present the argument in an abstract way, I'll go through it using the example of the cancer test that we just solved.

To keep the arithmetic simple, I'll assume a total population of exactly 10,000 people. Since all we are ultimately concerned about is percentages, this simplification will not affect the final answer. I'll also assume that the various probabilities are reflected exactly in the actual numbers. Thus, of the total population of 10,000, precisely 100 will have the cancer and 9,900 will not.

In the absence of the test, all you can say about the likelihood of your having the cancer is that there is a 1 percent chance that you do. This is the prior probability. Then you take the test, and it shows positive. How do you revise the probability that you have the cancer?

Well, there are 100 individuals in the population who have the cancer, and for all of them, the test will correctly give a positive prediction, thereby identifying 100 individuals as having the cancer.

Turning to the 9,900 cancer-free individuals, for 21 percent of them, the test will incorrectly give a positive result, thereby identifying $9,900 \times 0.21 = 2,079$ individuals, as having the cancer.

Thus, in all, the test identifies a total of $100 + 2,079 = 2,179$ individuals as having the cancer. Having tested positive, you are among that group. (This is precisely what the test evidence tells you.) The question is, Are you in the subgroup that really does have the cancer, or is your test result a false positive?

Of the 2,179 identified by the test, 100 really do have the cancer. Thus, the probability of your being among that group is $100/2,179 = 0.0459$.

And that's it.

Despite its simplicity, Bayes' method can be tricky to apply correctly, and if you are not careful, it is easy to go astray. The issue is not so much the formula but the concept of conditional probabilities.

For instance, unscrupulous lawyers have been known to take advantage of the innumeracy of judges and juries by deliberately confusing the two conditional probabilities $P(G|E)$, the probability that the defendant is guilty, given the evidence, and $P(E|G)$, the conditional probability that the evidence would be found, assuming the defendant were guilty. Such misuse of probabilities is called the *prosecutor's fallacy*, and even ethical prosecutors sometimes succumb to it unintentionally in cases where scientific evidence such as DNA testing is involved (e.g., paternity suits and rape and murder cases). Prosecuting attorneys in such cases have been known to provide the court with a figure for $P(E|G)$, the likelihood (more precisely, the *un*likelihood) of finding the evidence if the defendant were guilty, whereas the figure relevant to deciding guilt is $P(G|E)$, which, as Bayes' formula shows, is generally much lower than $P(E|G)$. Unless there is other evidence that puts the defendant into the group of possible suspects, this use of $P(E|G)$ is highly suspect and perhaps should be prohibited. The reason is that it ignores the initial low prior probability that a person chosen at random is guilty of the crime in question, and, as the example of the cancer test shows, the difference can be substantial.

Bayes' formula essentially took probability inside the human mind—applying it to our knowledge of a single, one-off event. With it, humankind had another powerful tool for predicting the future. But it took mathematicians many decades to discover just how powerful it can be.

CHAPTER 10

The Measure of Our Ignorance

Sunday, July 25, 1660

Monsieur,

As soon as I discovered that we were nearer to one another than we had ever been before, I could not resist making plans for renewing our friendship and I asked M. de Carcavi to be mediator: in a word I would like to embrace you and to talk to you for a few days; but as my health is not any better than yours, I very much hope that you will do me the favor of coming half way to meet me and that you will oblige me by suggesting a place between Clermont and Toulouse, where I would go without fail towards the end of September or the beginning of October.

If you do not agree to this arrangement, you will run the risk of seeing me at your house and of thus having two ill people there at once. I await your news with impatience and am, with all my heart,

Yours ever,
Fermat

The letter above could be called Fermat's last letter. It is the last one he is known to have written to his old correspondent Pascal. The two men whose remarkable exchange in the summer of 1654 had set in motion a revolution had never met, nor had they corresponded after they solved the problem of the points. But hearing six years later that Pascal was nearby, Fermat could not resist trying to arrange a meeting. Pascal replied immediately.

Tuesday, August 10, 1660

Monsieur,

You are the most gallant man in the world and assuredly I am the one who can best recognize your qualities and very much admire them, especially when they are combined with your own singular abilities. Because of this I feel I must show my appreciation of the offer you have made me, whatever difficulty I still have in reading and writing, but the honor you do me is so dear to me that I cannot hasten too much in answering your letter.

I will tell you then, Monsieur, that if I were in good health, I would have flown to Toulouse and I would not allow a man such as you to take one step for a man such as myself. I will tell you also that, even if you were the best Geometrician in the whole of Europe, it would not be that quality which would attract me to you, but it is your great liveliness and integrity in conversation that would bring me to see you.

For, to talk frankly with you about Geometry, is to me the very best intellectual exercise: but at the same time I recognize it to be so useless that I can find little difference between a man who is nothing else but a geometrician and a clever craftsman. Although I call it the best craft in the world, it is after all only a craft, and I have often said it is fine to try one's hand at it but not to devote all one's powers to it.

In other words, I would not take two steps for Geometry and I feel certain you are very much of the same mind. But as well as all this, my studies have taken me so far from this way of thinking, that I can scarcely remember that there is such a thing as geometry. I began it, a year or two ago, for a particular reason; having satisfied this, it is quite possible that I shall never think about it again.

Besides, my health is not yet very good, for I am so weak that I cannot walk without a stick nor ride a horse, I can only manage three or four leagues in a carriage. It was in this way that I took twenty-two days in coming here from Paris. The doctors recommended me to take the waters at Bourbon during the month of September, and two months ago I promised, if I can manage it, to go from there through Poitou by river to Saumur to stay until Christmas with M. le duc de Roannes, governor of Poitou, who has feelings for me that I do not deserve. But, since I go through Orléans on my way to Saumur by

*river and if my health prevents me from going further. I
shall go from there to Paris.*

*There, Monsieur, is the present state of my life, which
I felt obliged to describe to you so as to convince you of
the impossibility of my being able to receive the honor
you have so kindly offered me. I hope, with all my heart,
that one day I shall be able to acknowledge it to you or to
your children, to whom I am always devoted, having a
special regard for those who bear the name of the fore-
most man in the world.*

I am, etc.

Pascal

De Bienassis, 10th August, 1660

That brief exchange in the summer of 1660 was, as far as we
know, the last time those two great minds came into contact.
They never were able to meet in person. Pascal's health continued to deteriorate, and two years and nine days after sending the letter, he was dead. Fermat lived another three years.

One can only admire Pascal, who in his time was already
recognized as one of the greatest minds in history, for ac-
knowledging that when it came to mathematics, Fermat simply towered above him. *"I would not allow a man such as
you to take one step for a man such as myself."* This was not
false modesty.

It is clear from their correspondence that while Pascal
struggled the whole time, Fermat's deeply penetrating mind
cut right to the heart of the problem of the points. His analy-

sis provides a copybook example of how mathematics is supposed to work—cutting away all that is irrelevant and focusing on the key underlying logical issues. While Pascal's letters are long and discursive, Fermat's are short and to the point. Indeed, it appears that once he had solved the problem Pascal brought him, he had no further interest in the issue. In his letters to Pascal, he kept trying to change the topic to the problems of number theory that he much preferred.* Fortunately for humankind, Fermat's interest was aroused sufficiently for him to solve the problem of the unfinished game before his focus shifted.

The legacies the two men left are enormous. Disregarding everything else they did in their highly productive lives, the one mathematical breakthrough contained in their brief exchange of letters in 1654—the entire crux of which can be found in Pascal's letter of August 24—started humanity on the path toward scientific risk management and thereby changed human life forever.

Just how well we can benefit from their legacy was demonstrated dramatically in 2001.

IF ONLY

If only we had known in advance what the infamous 9/11 airplane hijackers were planning to do, we might have prevented them from achieving their murderous objective.

---

*See, for example, the letter reproduced at the end of Chapter 6.

That, at least, was a typical lament in the days and months that followed. As it happens, we did "know." A software system commissioned by the U.S. Department of Defense had alerted us to the risk a few months earlier. But nobody did anything about it. (This is not a question of anyone's dropping the proverbial ball—read on.)

In May 2001, a software system called Site Profiler was delivered to all U.S. military installations around the world. The system provided site commanders with tools to help assess terrorist risks so that they could develop appropriate countermeasures. It worked by combining different data sources to draw inferences about the risk of terrorism, making repeated use of Bayes' formula. Before the system was deployed, its developers carried out a number of simulations, based on hypothetical threat scenarios, which they referred to in a paper they wrote in 2000.* Summarizing the results of the tests, they noted, "While these scenarios showed that the RIN [Risk Influence Network] 'worked,' they tended to be exceptional (e.g. attacks against the Pentagon)."** In other words, the Pentagon was a prime terrorist target.

As we learned to our horror just a few months later, the Pentagon was one of the sites hit in the September 11 attack

---

*Linwood D. Hudson et al., "An Application of Bayesian Networks to Antiterrorism Risk Management for Military Planners," preprint. Subsequent to the events of September 11, 2001, the authors revised the paper several times. The original preprint version, while widely circulated, was never published in that form.

**Ibid., section 3.2.4.

on the United States. Unfortunately, though understandably, neither the military command nor the U.S. government had taken seriously Site Profiler's prediction that the Pentagon was in danger from a terrorist attack—nor for that matter had the system developers themselves, who viewed that particular prediction as "exceptional." The fact is that, before September 11, 2001, no one (other than the terrorists) regarded the Pentagon as a *serious* target of a terrorist attack. Yet when presented with the available data, Bayes' rule picked it right out. The mathematics was spot-on.

In fact, it was still something of a fluke that the Pentagon was the one example cited in the system developers' report. (It was also a fluke that the Pentagon was actually hit; as subsequent investigations discovered, it was a secondary target to be used if the airplane was unable to hit the White House.) From our present perspective, however, the real story is the degree to which the mathematics was able to provide a chillingly accurate assessment of a future risk. The math worked.

Site Profiler is just one of many systems based on Bayesian inference. Such systems allow users to estimate and manage a large range of risks by using Bayes' formula iteratively to combine evidence from different data sources: satellite imagery, message interception and decryption, human intelligence gathering, analytic models, simulations, historical data, and human judgments. The user of such a system typically enters information about, say, a military installation's assets through a question-and-answer interface reminiscent of a tax-preparation package. (The developers of Site Profiler actually modeled its interface on TurboTax.) The software uses this information to evaluate the various risks and finally outputs a list of threats, each one with a numerical rank based on its likelihood, the severity of its consequences, and other measures.

Essentially the same technique is used to produce the daily "terrorist threat" at airports around the world. The airport security personnel adjust the level of passenger screening according to the calculated "threat level." This may be stated in the form of a color code, with red the highest level, then orange, and so on, or else using words such as "critical," "severe," and so forth. In either case, the threat categories are based on numerical ranges, with the underlying numbers coming from an iterated use of Bayes' formula. Each day, as new information comes in, Bayes' formula is used to revise the probability figure from the previous day.

The point is not that such systems eliminate risk; that is impossible. They reduce it. Whereas people can often find good reasons to ignore or dismiss a particular risk, Bayes' theorem simply grinds the numbers and gives a precise answer. Since 9/11, the authorities have been taking those answers much more seriously. Although hard evidence is not available—because the government understandably keeps it secret—there are reasons to believe that a number of terrorist attacks have been thwarted, in part because the use of Bayes' rule narrowed down the list of imminent threats sufficiently to initiate countermeasures.

MEASURING OUR IGNORANCE

Events such as the September 11 terrorist attack are, thankfully, extremely rare. One consequence is that when we apply the methods of probability theory to compute the likelihood of such an event, our problem is very different from what we find at the gaming table or in the life insurance company office, where the same kind of event (a double 6 or an automobile accident) occurs often.

The stakes are different, too. When a customer wins a large sum in the casino or the life insurance company sells a policy to a person who dies the next day, the consequences for the casino or insurance company are essentially irrelevant. The loss is made up by the very large number of casino gamblers who put up their stake and don't win, or the tens of thousands of insurance policyholders who pay their premiums and don't die. But a *single* event like 9/11 is a disaster. It is truly remarkable that the same mathematical notion of probability can be applied in both kinds of situations. And make no mistake: although there are several ways to interpret and apply probabilities, there is just one underlying *mathematical* notion.

When we use probability to manage the risk associated with a single, essentially unrepeatable event, what we are really doing is quantifying our ignorance. We may be ignorant because the event in question is in the future, such as the next terrorist attack. Alternatively, the event may have taken place, and our ignorance is due simply to having insufficient information.

For an example of the latter, if I tell you that I have two children, and this is all you know about my family, then, if you are acting rationally, you would assign the following probabilities to the composition of my family:*

Probability of two boys   =   0.25

Probability of two girls   =   0.25

Probability of one boy and one girl   =   0.5

If I offered you an even-money bet on which of the three possibilities is the case, you would be wise to bet on my having one child of each sex.

What the above probabilities measure, however, is *your lack of information about my family.* It happens that I have two daughters, so *for me* the probabilities are:

Probability of two boys   =   0

Probability of two girls   =   1

Probability of one boy and one girl   =   0

For cognitive reasons that are not fully understood, while our intuitions regarding the *a priori* probabilities that arise in games of chance are fairly good, we are easily misled when we try to use probability to quantify our knowledge. For example, suppose you meet my brother Peter and he says, "I have two children, and at least one of them is a girl."

---

*We assume the probability of a male or female child at each birth is 50 percent.

What would you judge to be the probability that my brother has a boy and a girl?

The most common answer is 50 percent. People typically reason like this: Peter's other child is equally likely to be a boy or a girl, so the probability that he has a boy and a girl is 50 percent. Unfortunately, this reasoning is not correct. In fact, Peter is *twice* as likely to have a boy and a girl as he is to have two girls.

The easiest way to understand why this is the case is to tackle the problem using the same approach that Fermat used with the problem of the unfinished game—namely, to list all the possibilities and simply count. Listed in order of births, there are four possibilities for Peter's family: B–B, G–G, B–G, G–B. As far as you are concerned, each of these is equally likely, so you assign each the probability 0.25.

When Peter tells you that at least one of his children is a girl, he eliminates the first possibility, leaving the following three: G–G, B–G, G–B. In two of these three possibilities, Peter has children of different genders, so you (should) calculate the probability of his having children of both sexes to be 2/3 and the probability of his having two girls as 1/3. Thus, he is (from your perspective) twice as likely to have a boy and a girl as he is to have two girls.

Are you convinced? If you think I should not have listed the two mixed-gender possibilities as two separate events (B–G and G–B), you are making the same mistake Pascal did with the problem of the unfinished game!

Probability Goes Back to Court

Another example that highlights our poor intuitions regarding probabilities is a surprising result known as the birthday paradox. There is no paradox, just a result that runs counter to our intuitions. The question is, How many randomly selected people do you need to assemble in a room for there to be a better-than-even chance that two of them will have the same birthday?

The most common answer is 183, just over one-half the number of days in a year. The correct answer is 23. The exact calculation is a bit intricate, but you get a sense of why the answer is so low when you realize that with 23 people, there are $23 \times 22 = 506$ possible pairs, each of which might share a birthday, and this turns out to be just enough pairs to tilt the odds to 0.508 in favor of there being a match.

The birthday paradox phenomenon turns out to be hugely important in courts of law these days, in connection with DNA identification, where identification of an individual is determined by comparing a DNA sample taken from the person with a sample taken from a reference source—for example, a crime scene.

Many people think the comparison is between the entire DNA sequences, but it isn't. What are compared are so-called DNA profiles. In DNA profiling, a number (previously eight, currently thirteen) of specific gene sequences are chosen as comparison loci, and a numerical marker (based on the molecular structure of that sequence) is calculated for each. The resulting sequence of numbers (so, thirteen of them in current DNA profiling) is the individual's DNA profile. It is the DNA profiles—those sequences of numbers—that are compared for DNA (profile) identification. When the scientists who do the comparison declare a match, they mean that the two DNA samples have the same DNA profiles.

Each of the thirteen comparison loci is chosen so that it is moderately unlikely that two unrelated people selected at random will have the same numerical marker at that locus. The probability of this occurring is about 1/10. Because the different loci are believed to be independent of one another (in the same way that two rolls of a die are independent), the probability of two randomly selected people having the same profile can be computed by multiplying the probabilities of a match at each locus. For thirteen loci, this comes to $1/10^{13}$, or 1 in 10 trillion. It is this tiny probability of an accidental match, where two people share the same profile, that gives DNA profiling its high degree of reliability.

But the method has its dangers. First, if two people are related, they are much more likely to have agreement on some loci, and the closer the relationship, the greater the agreement may be. This can rapidly erode those impressive odds. Second, the birthday paradox means that in a large city of several million people, there is actually a real chance of two people having the same profile. In fact, even in a small town of just sixty-five thousand people, the probability of two people selected at random having DNA profiles that agree on as many as nine loci is around 5 percent. (This is why the procedure was modified from eight-locus profiles to thirteen loci, although with the ability of today's databases to store millions of profiles, even thirteen-locus profiles are starting to look inadequate.*)

The courts, you will be relieved to know, are aware of this issue (at least the better-informed ones are), and in major cases, they will not accept DNA profile evidence without an estimate from professional statisticians regarding the likelihood that a particular DNA profile match is accidental. This means, however, that judges and juries are sometimes faced with having to grasp the full implications of the numbers those experts submit, and the growing list of cases now before various appeals courts suggests that they often get it wrong.

Pascal himself, to judge from his letters to Fermat in the summer of 1654, would likely have had difficulty following the calculations. Fermat, on the other hand (whose day job was as a jurist, remember), almost certainly would have had no problem.

Where even Fermat would have difficulty, along with everyone else, is in saying exactly what a probability figure is.

---

*The inadequacy really manifests only in so-called cold-hit cases, where a crime suspect is found by means of a search through a database of DNA profiles. As a means of court identification of someone found *by means other than DNA*, the method is the surest form of identification there is.

WHAT EXACTLY IS PROBABILITY?

I have a friend named Ed who e-mails me now and then to comment on articles I have written in my online column "Devlin's Angle." Whenever I write about probability theory, as I do from time to time, Ed generally sends me a message challenging what I say, claiming that the notion of probability I discuss makes no sense. Ed has a Ph.D. in mathematics, and I got to know him when we were both setting out on academic careers, but Ed left university life early on to pursue a successful career in the financial markets. He understands the issues full well. He also appreciates my standpoint, though he steadfastly maintains that he does not. The fact is, he has a point. When you sit down and really try to understand what a numerical probability is, you find you cannot. Philosophers have written entire books trying to come to terms with the concept.

The first thing that is problematic is that a probability is assigned to a single event, not to the repetition of the action. In the case of rolling a die, the 0.5 probability that the outcome will be even is a feature of the action of rolling the die (once). It tells you something about how that single action is likely to turn out. It does not, of course, mean that the die will land showing half even and half odd. One or the other will occur. The 0.5 probability is something that arises over many repetitions, and only by repeating the action many times are you likely to observe the pattern of outcomes that the probability figure captures.

Still, for all its strangeness, the kind of probability you en-counter when rolling dice or tossing coins does not cause many people much trouble—apart perhaps from calculating the odds in all but the simplest cases.

It is true that we often compute such probabilities by merely *imagining* what would occur if we were to repeat the action many times, basing our reasoning on symmetry con-siderations (the coin is symmetrical, each face of the die is the same as any other, apart from the numbers it shows, etc.). Yet probability is an empirical notion. You *can* test it by experiment.

My friend Ed has no problem with this kind of probabil-ity. He starts to object when probabilities are attached to people's knowledge of an event, particularly a one-off event, such as the puzzle I presented earlier about my brother Peter's two children. After Peter tells you that he has two children and that at least one of them is a girl, you may cor-rectly reason that, from your perspective, the probability that he has two girls is 1/3. Yet you know that one of two things is true: either he has two daughters, or he has a son and a daughter. If you acquire additional information, you can revise your probability figure. For instance, if he tells you that his *elder* child is a daughter, you would (if you rea-soned correctly) assign 1/2 as the probability that he has two daughters. His giving you additional information has not changed his family, of course, only your knowledge about it. This is different from what happens when we assign proba-bilities to outcomes of dice rolls or coin tosses.

It is this second kind of probability, known as *epistemic probability,* that Ed objects to. "It makes no sense to assign numbers to our knowledge in this way," he maintains. And this is before we get to highly subjective examples, such as the person who says, "I'm 95 percent certain I turned the gas off before I left," or mixtures of objective calculation and human judgment, such as when the television weatherman tells us there is a "30 percent chance of rain."

We usually try to explain away such references to probabilities by saying something like "If the situation were to repeat itself many times, then in N percent of the cases, such and such would happen." But there cannot be such repetition; it is a one-off event. In the case of the weather, you can argue differently by saying that when you look back over the historical record, you find that of all those days when the conditions were more or less as they are today, then in 30 percent of those cases, rain followed. The weather forecast may be reliable, but as Ed would remind us, there never are two days when all conditions are the same, so the 30 percent figure is not mathematically precise—it is not a *probability.* The weather is not the same as rolling dice.

Still, most of us regularly base our actions on such figures. (I'm sure Ed does, too.) If you know your friend is reliable and she says she is "95 percent certain" she turned off the gas, you most likely will not go back and check, whereas if she says she is "only 30 percent certain," you surely will. Similarly, faced with a "75 percent chance of rain," you are likely to postpone the barbecue, whereas a

"10 percent chance of rain" is unlikely to make you change your plans.

How can a mathematician make sense of such numbers? This question fascinated an Italian mathematician called Bruno de Finetti (1906–1985). He suggested a way to interpret even highly subjective probability assignments like your friend's 95 percent certainty that she turned off the gas. Of course, she may be saying "95 percent certain" in a purely figurative way to mean "I'm very certain." The actual number one uses can be arbitrary, so long as it is fairly close to 100. But you can, de Finetti suggested, provide a rationale for assigning a number. The idea is to play what is today called a *de Finetti game*.

Suppose you are the person who makes the claim (that you are 95 percent certain you turned off the gas). I now offer you a deal. I present you with a jar containing 100 balls, 95 of them red, 5 black. You have a choice. You can draw one ball from the jar, and if it's red, you win $1 million. Or we can go back and see if the gas is on, and if it is not, I give you $1 million.

Now, if your "95 percent certain" claim were an accurate assessment of your confidence, it would not make any difference whether you choose to pick a ball from the jar or go back with me and check the status of the gas stove. But I suspect that when it comes to the crunch, you will elect to pick a ball from the jar. After all, there is only 1 chance in 20 that you will fail to win $1 million. You'd be crazy not to go for it.

By electing to pick a ball, you have demonstrated that what I will call your *rational confidence* that you have turned off the gas is at most 95 percent.

Now I offer you a jar that contains 90 red balls and 10 black, with choosing a red ball again netting you $1 million. If you again choose to select a ball rather than go and check your stove, we may conclude that your rational confidence that you have turned off the gas is at most 90 percent. If it were more than that, you should decline the ball-picking offer and take me to your house to check your stove. (So much for your "95 percent" claim!)

Then I offer a jar with 80 red balls and 20 black. If you choose to pick a ball this time, your rational confidence that you have turned off the gas can be at most 80 percent.

And so on.

Eventually, of course, you decide you would rather check your stove than select a ball from the jar. If that happens when there are N red balls in the jar, then your rational confidence is precisely N percent. The de Finetti procedure has established an exact correspondence between your subjective probability and a frequentist probability (namely, picking balls from a jar where you know the proportions of the colors of the balls).

De Finetti did not suggest that his procedure would yield an objective answer. Such a game clearly has a huge psychological aspect, and on different occasions, you might settle on different values for when to go back and check the gas, depending on your mood. He did not propose his game as a practical procedure. Rather, his purpose was to suggest a hypothetical procedure to give a *meaning* to the number you quote—an answer to the question "Where does the number come from?" (or more likely, "What number should you use to justifiably reflect your subjective belief?").

THE FORESIGHT SAGA

The Pascal-Fermat correspondence showed that it is possible to use mathematics to see into the future. The foresight you get from probability theory is, of course, a probabilistic one. You do not know exactly what is going to happen, but you can compute numerical likelihoods. It turned out that probability theory is not the only kind of mathematics that can provide us with a sort of foresight that we may use to our advantage. Once mathematicians realized that mathematics can provide foresight, as they did following Pascal-Fermat, it was not long before they started finding other ways to make such use of mathematics.

Just as the original Pascal-Fermat work arose from gamblers' desire to improve their chances at the gaming tables, many of the other developments of "foresight mathematics" were also motivated by financial gain—for example, in the financial markets. Today's investment brokers make extensive use of a range of mathematical techniques to increase their chances of making a profit. I'll describe one by way of an illustration.

The 1997 Nobel Prize for economics went to the two surviving members of a three-person team that developed a mathematical formula for predicting the future behavior of a particular class of investment objects known as market derivatives. At the time of the award, Myron Scholes was an emeritus professor of finance at Stanford University and Robert C. Merton was a professor of economics at Harvard University. Together with a third man, Fischer Black, who died in 1995, they revolutionized the financial markets with their discovery of what is known as the Black-Scholes formula. (It was discovered by Scholes and Black and developed by Merton.) This formula tells investors what value to put on a financial derivative, such as a stock option. By turning what had previously been a guessing game into a mathematical science, it made the derivatives market the hugely lucrative industry it is today.

The very idea that you could use mathematics to price derivatives was so revolutionary that Black and Scholes initially had difficulty publishing their work. When they first tried, in 1970, the University of Chicago's *Journal of Political Economy* and Harvard's *Review of Economics and Statistics* both rejected the paper without even bothering to have it refereed. It was only in 1973, after some influential members of the Chicago faculty put pressure on the journal editors, that the *Journal of Political Economy* published the paper.

Industry was far less shortsighted (perhaps a better term would be *blind-sighted*) than the ivory-tower journal editors. Within six months of the publication of the Black-Scholes article, Texas Instruments had incorporated the new formula into its latest calculator and announced the new feature with a half-page advertisement in the *Wall Street Journal*.

The idea of a stock option is that you purchase an option to buy stock at an agreed price prior to some fixed later date. If the value of the stock rises above the agreed price before the option runs out, you buy the stock at the agreed (lower) price. If you want, you can sell the stock immediately and realize your profit. If the stock does not rise above the agreed price, you don't have to buy it, but you lose the money you used to purchase the option.

What makes stock options attractive is that the purchaser knows in advance what the maximum loss is: the cost of the option. The potential profit is theoretically limitless: if the stock rises dramatically before the option runs out, you stand to make a killing. Stock options are particularly attractive in a market that sees large, rapid fluctuations, such as the computer and software industries. Most of the many thousands of Silicon Valley millionaires became rich because they elected to take a portion of their salary as stock options in their new company.

The question is, What is a fair price to charge for an option on a particular stock? This is precisely the question that Scholes, Black, and Merton investigated in the late 1960s. Black, a mathematical physicist with a recent doctorate from Harvard, had left physics to work for Arthur D. Little, the Boston-based management consulting firm. Scholes had just obtained a Ph.D. in finance from the University of Chicago.

Merton had a bachelor of science degree in mathematical engineering at Columbia University and was working at MIT as a teaching assistant in economics.

The three young researchers—all still in their twenties—set about trying to find an answer using mathematics. In so doing, they were very definitely going against the received wisdom of the time, which said you could not use mathematics to predict the market behavior of a derivative. Many senior traders—perhaps ignorant of the Pascal-Fermat correspondence—thought options trading was entirely a wild gamble, strictly for the foolhardy. On the other hand, that received wisdom was based more on a belief than evidence, since options trading was just being developed at the time. (The Chicago Board Options Exchange opened in April 1973, just one month before the Black-Scholes paper appeared in print.)

The old guard turned out to be wrong. Mathematics could be applied. It was heavy-duty mathematics, involving an obscure technique known as stochastic differential equations. The formula takes four input variables—duration of the option, prices, interest rates, and market volatility—and produces a price that should be charged for the option. It involves probabilities—that's what the word *stochastic* means. But the underlying mathematics originates with Daniel Bernoulli's efforts to apply techniques of calculus to the flow of liquids and gases. In broad terms, what Black and Scholes did was think of the flow of money between investors and financial institutions as if it were water flowing along a system of rivers. Just as the differential equations developed by Bernoulli and his successors enable you to say with precision exactly where the fluid will flow, how much, and how fast, so too the Black-Scholes formula allowed you to say, with *stochastic precision*, how and where the value of a derivative would "flow."

Not only did this audacious idea work, but it transformed the market. When the Chicago Board Options Exchange opened in 1973, fewer than a thousand options were traded on the first day. By 1995, over a million options were changing hands daily.

In fact, so great was the role played by the Black-Scholes formula (and extensions due to Merton) in the growth of the new options market that when the American stock market crashed in 1978, the business magazine *Forbes* put the blame squarely on that one formula. Scholes himself has said that it was not so much the formula that was to blame, but rather that market traders were not sophisticated enough about its use.

Because of the way markets work, once *everyone* started to use the Black-Scholes formula, much of the advantage it provided disappeared. Today, the Black-Scholes formula is just one of several mathematical formulas investors use to guide their decisions. It would be an unwise reader who, after completing this book, got hold of a Black-Scholes software package and sought to make a killing in derivatives. Like the other mathematical tools we now use, the Black-

Scholes formula does not eliminate uncertainty; it merely helps us to judge the risks.

What would have been unthinkable before Pascal and Fermat solved the problem of the unfinished game is that we may quantify those judgments, calculating precise numerical values for what tomorrow may bring. That is the legacy of the Pascal-Fermat correspondence. We may not know what tomorrow *will* bring, but we can quantify what it *might* bring, and act accordingly.

That is how we all live our lives.

THE KEY LETTER
FROM PASCAL TO FERMAT

Monday, August 24, 1654

Monsieur,

   1. I was not able to tell you my entire thoughts regarding the problem of the points by the last post, and at the same time, I have a certain reluctance at doing it for fear lest this admirable harmony which obtains between us and which is so dear to me should begin to flag, for I am afraid that we may have different opinions on this subject. I wish to lay my whole reasoning before you, and to have you do me the favor to set me straight if I am in error or to indorse me if I am correct. I ask you this in all faith and sincerity for I am not certain even that you will be on my side.

   When there are but two players, your theory which proceeds by combinations is very just. But when there are three, I believe I have a proof that it is unjust that you should proceed in any other manner than the one I have. But the method which I have disclosed to you and which I have used universally is common to all imaginable conditions of all distributions of points, in the place of that of combinations (which I do not use except in partic-ular cases when it is shorter than the general method), a method which is good only in isolated cases and not good for others.

I am sure that I can make it understood, but it re-quires a few words from me and a little patience from you.

2. This is the method of procedure when there are two players. If two players, playing in several throws, find themselves in such a state that the first lacks two points and the second three of gaining the stake, you say it is necessary to see in how many points the game will be ab-solutely decided.

It is convenient to suppose that this will he in four points, from which you conclude that it is necessary to see how many ways the four points may be distributed between the two players and to see how many combina-tions there are to make the first win and how many to make the second win, and to divide the stake according to that proportion. I could scarcely understand this reason-ing if I had not known it myself before; but you also have written it in your discussion. Then to see how many ways four points may be distributed between two players, it is necessary to imagine that they play with dice with two faces (since there are but two players), as heads and tails, and that they throw four of these dice (because they play in four throws). Now it is necessary to see how many ways these dice may fall. That is easy to calculate. There can be sixteen, *which is the second power of* four; *that is to say, the square. Now imagine that one of the faces is marked* a, *favorable to the first player. And suppose the other is marked* b, *favorable to the second. Then these four dice can fall according to one of these sixteen arrangements.*

```
a a a a a a a a b b b b b b b b
a a a a b b b b a a a a b b b b
a a b b a a b b a a b b a a b b
a b a b a b a b a b a b a b a b
1 1 1 1 1 1 2 1 1 1 2 1 2 2 2
```

and, because the first player lacks two points, all the arrangements that have two a's *make him win. There are therefore 11 of these for him. And because the second lacks three points, all the arrangements that have three* b's *make him win. There are 5 of these. Therefore it is necessary that they divide the wager as 11 is to 5.*

There is your method, when there are two players, whereupon you say that if there are more players. it will not be difficult to make the division by this method.

3. On this point, Monsieur, I tell you that this division for the two players founded on combinations is very equitable and good, but that if there are more than two players, it is not always just and I shall tell you the reason for this difference. I communicated your method to [some of] our gentlemen, on which M. de Roberval made me this objection:

That it is wrong to base the method of division on the supposition that they are playing in four *throws seeing that when one lacks* two *points and the other* three, *there is no necessity that they play* four *throws since it may happen that they play but* two *or* three, *or in truth perhaps* four.

Since he does not see why one should pretend to make a just division on the assumed condition that one plays four throws, in view of the fact that the natural terms of the game are that they do not throw the dice after one of the players has won; and that at least if this is not false, it should be proved. Consequently he suspects that we have committed a paralogism.

I replied to him that I did not found my reasoning so much on this method of combinations, which in truth is not in place on this occasion, as on my universal method from which nothing escapes and which carries its proof with itself. This finds precisely the same division as does the method of combinations. Furthermore, I showed him the truth of the divisions between two players by combinations in this way. Is it not true that if two gamblers finding according to the conditions of the hypothesis that one lacks two *points and the other* three, *mutually agree that they shall play four complete plays, that is to say, that they shall throw four two-faced dice*

*all at once—is it not true, I say, that if they are pre-
vented from playing the four throws, the division should
be as we have said according to the combinations favorable to each? He agreed with this and this is indeed
proved. But he denied that the same thing follows when
they are not obliged to play the four throws. I therefore
replied as follows:*

*It is not clear that the same gamblers, not being con-
strained to play the four throws, but wishing to quit the
game before one of them has attained his score, can without loss or gain be obliged to play the whole four plays,
and that this agreement in no way changes their condition? For if the first gains the two first points of four, will
he who has won refuse to play two throws more, seeing
that if he wins he will not win more and if he loses he will
not win less? For the two points which the other wins are
not sufficient for him since he lacks three, and there are
not enough [points] in four throws for each to make the
number which he lacks.*

*It certainly is convenient to consider that it is ab-
solutely equal and indifferent to each whether they play
in the natural way of the game, which is to finish as soon
as one has his score, or whether they play the entire four
throws. Therefore, since these two conditions are equal
and indifferent, the division should he alike for each. But
since it is just when they are obliged to play the four
throws as I have shown, it is therefore just also in the
other case.*

That is the way I prove it, and, as you recollect, this proof is based on the equality of the two conditions true and assumed in regard to the two gamblers, the division is the same in each of the methods, and if one gains or loses by one method, he will gain or lose by the other, and the two will always have the same accounting.

4. Let us follow the same argument for three *players and let us assume that the first lacks* one *point, the second* two, *and the third* two. *To make the division, following the same method of combinations, it is necessary to first discover in how many points the game may he decided as we did when there were two players. This will be in three points for they cannot play three throws without necessarily arriving at a decision.*

It is now necessary to see how many ways three throws may he combined among three players and how many are favorable to the first, how many to the second, and how many to the third, and to follow this proportion in distributing the wager as we did in the hypothesis of the two gamblers.

It is easy to see how many combinations there are in all. This is the third power of 3; that is to say, its cube, or 27. For if one throws three dice at a time (for it is necessary to throw three times), these dice having three faces each (since there are three players), one marked a *favorable to the first, one marked* b *favorable to the second, and one marked* c *favorable to the third,—it is evident that these three dice thrown together can fall in 27 different ways as:*

aaa	aaa	aaa	bbb	bbb	bbb	ccc	ccc	ccc
aaa	bbb	ccc	aaa	bbb	ccc	aaa	bbb	ccc
abc	abc	abc	abc	abc	abc	abc	abc	abc
111	111	111	111	1	1	111	1	1
	2			2	222	2		2
		3				3	3	3 333

Since the first lacks but one point, then all the ways in which there is one a *are favorable to him. There are 19 of these. The second lacks two points. Thus all the arrangements in which there are two* b*'s are in his favor. There are 7 of them. The third lacks two points. Thus all the arrangements in which there are two* c*'s are favorable to him. There are 7 of these. If we conclude from this that it is necessary to give each according to the proportion 19, 7, 7, we are making a serious mistake and I would hesitate to believe that you would do this. There are several cases favorable to both the first and the second, as* abb *has the* a *which the first needs, and the two* b*'s which the second needs. So too, the* acc *is favorable to the first and third.*

It therefore is not desirable to count the arrangements which are common to the two as being worth the whole wager to each, but only as being half a point. For if the arrangement acc *occurs, the first and third will have the same right to the wager, each making their score. They should therefore divide the wager in half. If the arrangement* aab *occurs, the first alone wins. It is necessary to make this assumption.*

There are 13 arrangements which give the entire wager to the first, and 6 which give him half and 8 which are worth nothing to him. Therefore if the entire sum is one pistole, there are 13 arrangements which are each worth one pistole to him, there are 6 that are each worth 1/2 a pistole, and 8 that are worth nothing.

Then in this case of division, it is necessary to multiply

	13	*by one pistole which makes*	13
	6	*by one half which makes*	3
	8	*by zero which makes*	0
Total	27		*Total* 16

and to divide the sum of the values 16 by the sum of the arrangements 27, which makes the fraction 16/27 and it is this amount which belongs to the first gambler in the event of a division; that is to say, 16 pistoles out of 27.

The shares of the second and the third gamblers will be the same:

There are	4	*arrangements which are worth 1 pistole; multiplying,*	4
There are	3	*arrangements which are worth 3/2 pistole; multiplying,*	1½
And	20	*arrangements which are worth nothing*	0
Total	27		*Total* 5½

Therefore 5 pistoles belong to the second player out of 27, and the same to the third. The sum of the 5 1/2, 5 1/2, and 16 makes 27.

5. It seems to me that this is the way in which it is necessary to make the division by combinations according to your method, unless you have something else on the subject which I do not know. But if I am not mistaken, this division is unjust.

The reason is that we are making a false supposition,—that is, that they are playing three throws without exception, instead of the natural condition of this game which is that they shall not play except up to the time when one of the players has attained the number of points which he lacks, in which case the game ceases.

It is not that it may not happen that they will play three times, but it may happen that they will play once or twice and not need to play again.

But, you will say, why is it possible to make the same assumption in this case as was made in the case of the two players? Here is the reason: In the true condition [of the game] between three players, only one can win, for by the terms of the game it will terminate when one [of the players] has won. But under the assumed conditions, two may attain the number of their points, since the first may gain the one point he lacks and one of the others may gain the two points which he lacks, since they will have played only three throws. When there are only two players, the assumed conditions and the true conditions concur to the advantage of both. It is this that makes the greatest difference between the assumed conditions and the true ones.

If the players, finding themselves in the state given in the hypothesis,—that is to say, if the first lacks one point, the second two, and the third two; and if they now mutually agree and concur in the stipulation that they will play three *complete throws; and if he who makes the points which he lacks will take the entire sum if he is the only one who attains the points; or if two should attain them that they shall share equally, in this case, the division should be made as I give it here. the first shall have 16, the second 5 1/2, and the third 5 1/2 out of 27 pistoles, and this carries with it its own proof on the assumption of the above condition.*

But if they play simply on the condition that they will not necessarily play three throws, but that they will only play until one of them shall have attained his points, and that then the play shall cease without giving another the opportunity of reaching his score, then 17 pistoles should belong to the first, 5 to the second, and 5 to the third, out of 27. And this is found by my general method which also determines that, under the proceeding condition, the first should have 16, the second 5 1/2, and the third without making use of combinations,—for this works in all cases and without any obstacle.

6. These, Monsieur, are my reflections on this topic on which I have no advantage over you except that of having meditated on it longer, but this is of little [advantage to me] from your point of view since your first glance is more penetrating than are my prolonged endeavors.

I shall not allow myself to disclose to you my reasons for looking forward to your opinions. I believe you have recognized from this that the theory of combinations is good for the case of two players by accident, as it is also sometimes good in the case of three gamblers, as when one lacks one point, another one, and the other two because, in this case, the number of points in which the game is finished is not enough to allow two to win, but it is not a general method and it is good only in the case where it is necessary to play exactly a certain number of times.

Consequently, as you did not have my method when you sent me the division among several gamblers, but [since you had] only that of combinations, I fear that we hold different views on the subject.

I beg you to inform me how you would proceed in your research on this problem. I shall receive your reply with respect and joy, even if your opinions should be contrary to mine.

I am etc.

INDEX

Basic Ideas

Every great idea—whether embodied in a speech, a mathematical equation, a song, or a work of art—has an origin, a birth, and a life of enduring influence. In each book in the Basic Ideas series, a leading authority offers a concise biography of a text that transformed its world, and ours.